THE TRAINED EYE

AN INTRODUCTION TO ASTRONOMICAL OBSERVING

LEON PALMER
El Camino College

Saunders College Publishing
Philadelphia Fort Worth Chicago
San Francisco Montreal Toronto
London Sydney Tokyo

Printed in the United States of America.

Palmer: THE TRAINED EYE: AN INTRODUCTION TO ASTRONOMICAL OBSERVING

ISBN # 0-03-047363-2

012 002 987654321

Dedication

With love to Linda for her collaboration and

to Kimberly and Matthew for their futures.

Instructor's Guide & Preface

The mediocre teacher tells,
the good one explains,
the superior one shows,
the great one inspires.
Anonymous

The Trained Eye laboratory manual is designed to serve five important instructional needs of an introductory college astronomy curriculum for non-majors:

- Provide hands-on experience to enrich understanding of important concepts covered in an astronomy survey/lecture class.
- Provide a complete course structure for an introductory astronomy laboratory class.
- Inspire the non-major to a lifelong interest in astronomy.
- Develop the non-major student's skills to pursue that interest.
- Guide the non-major student beyond his/her formal instruction in astronomy.

These goals have driven the pedagogical organization of the text and projects that compose the Trained Eye laboratory manual and its accompanying volume, The Trained Eye Star Atlas.

The Text

The text is written in a readable and personable style directed to the student. The chapters develop astronomical concepts in a natural flow; the framework of the celestial sphere, how the celestial sphere relates to the horizon coordinnate system, the principles of telescope optics and mountings, collecting astronomical observations, the informational content of the observations, pre-observing planning and advanced observing techniques.

The text presents the student's innate skills of perception, visual pattern recognition, perception of brightness, attention, the physiology of sight, magnitudes, night vision, color vision, andaverted vision. The projects make astronomical concepts real and develop the student's skills of perception into skills of astronomical observation.

The Projects

The projects are divided between cloudy night (indoors) and clear night (outdoors) and are printed in a larger typeface than the text . Whether cloudy or clear, all projects bring life to concepts and meaning to observation by presenting concepts through their observables. For example, rather than having the student measure dots (star images) on a piece of paper, the student learns to *see* magnitudes by visually estimating the magnitudes of stars from their images upon a projected slide of a star cluster. The student also learns how to use his/her innate skill of pattern matching to construct the mental star map that connects the finding chart to the star field — an innate skill developed into an observing skill. Two projects use slides in this way (Projects 1-1 and 5-1). The slides used are, for the most part, ones already in the instructor's slide collection. In all cases the slides are readily available at nominal cost from sources indentified in those projects.

Most of the cloudy night exercises can be completed in 1 1/2 hours, some require a 3 hour laboratory class to perform completely. These longer projects have been divided into <1 hour parts to give the instructor flexibility in fitting all or part of these longer cloudy-night projects into his/her class time limitations. All the clear-night projects in the main text require a minimum of 3 hours of observation. They are intentionally open-ended to permit the instructor (or the student) flexibility. The projects in Appendix A are intended as supplemental to the projects in the chapters and follow no particular order. Several require a total of 2 hours of observation spread over one or more months. Others require 3 to 6 hours of observations spread over a night or several nights. These time requirements are flexible and left to the instructor. The instructor may also wish to supplement a limited series of observations with data from other sources (e.g., star charts, the Astronomical Almanac).

The Star Atlas

The use of The Trained Eye Star Atlas is intimately woven throughout the text and projects. It is a tool, equal in importance to the telescope, that a student will use long after completing the course. The Trained Eye Star Atlas provides the right depth of detail and yet simplicity of design to span the needs of the novice to the professional.

Notes on Individual Projects

Project 1-1 Charts, Stars and Magnitudes — cloudy night

The student learns to see magnitudes by visually estimating the magnitudes of starfrom their images upon a projected slide of the star cluster M67. The slide is taken in the V filter and represents visual magnitudes, while the finding chart is from a B filter photograph and represents blue magnitudes. The slide can be obtained from:

Hansen Planetarium
15 S. Lake St.,
Salt Lake City, Utah 84111
(801) 534-2104

By comparing images for several stars on both the slide and the finding chart, the student is introduced to colors and how color can be recorded in black and white photographs. The student also learns how to use his/her innate skill of pattern matching to construct the mental star map that connects the finding chart to the star field and identify problems that affect the estimation of magnitudes. The slide of M67

Project 1-2 Mapping the Heavens — cloudy night

This project teaches usage of a star chart by presenting a number of stars to find in the star atlas, some to be found from their celestial coordinnates and others from their Bayer-Flamsteed name. The concept of magnitudes is reinforced by having the student compile statistics on the brightest stars and converting magnitudes to brightness ratios. When the student is asked to measure the approximate distances between stars on the atlas, he/she may use the declination scale on the star charts to convert distance in mm to degrees. Because of the projection used in The Trained Eye Star Atlas, the error introduced in measuring arc distances in either right ascension or declination is minor.

Project 2-1 The Night Sky — clear night

This project teaches the measurement of altitudes, azimuths, colors of stars, how the celestial sphere appears in the horizon coordinnate system, how to find stars in the sky from stars on a chart, and how to use the star chart to predict the location, in horizon coordinnates, of stars that aren't up. All measurements are made with the student's hand held up at arms length (approximately 10°). With care, an accuracy of ±5° is achievable.

Project 2-2 Diurnal Motions — clear night or cloudy night (planetarium)

In this project, the student sees the diurnal motions of stars as a function of declination, learns how to determine where and when stars rose or will set, and how local sidereal time relates to right ascension and the meridian, and how to use local sidereal time to predict the location of other points on the celestial sphere in horizon coordinnates. Measurements are made by hand as in Project 2-1. Observations should be spread over at least 3 hours if the motions are to be clearly seen.

Project 3-1 Lenses and Telescopes — cloudy night

This project teaches the concepts covered in the chapter: objective lens, eyepiece lens, focus, image scale, spherical aberration, chromatic aberration, magnification, field of view, resolution, f-ratio and their interdependencies. The topics are grouped into five parts for convenience and summary questions at the end of the project are provided to establish relationships between different properties (e.g. magnification and field of view). The equipment requirements are summarized at the start of the project. The instructor can adapt this project to his/her equipment.

Project 4-1 Telescope Field of View and Light-Gathering Power — clear night

In this project, the student measures field of view and light-gathering power for the unaided eye, finder telescope, and main telescope. Field of view is measured using drift times of stars near the celestial equator. Light gathering power is determined by estimating the faintest stars that can be seen visually in a constellation with the unaided eye and telescopically in either the Pleaides or Praesepe star clusters. By making unaided eye observations at different altitudes, the student can qualitatively measure the effects of atmospheric absorption. For the finder telescope, where identification of individual stars is difficult, a statistical approach based upon counting the number of stars in the cluster is described. The effects of magnification on image contrast and the ability to see faint stars are also explored.

Project 4-2 Visual Photometry of Stars — clear night

This project is identical to Project 1-1 except that the student makes actual observations of stars through the telescope to estimate visual magnitudes and develop his/her observing skills. Stars in the Pleaides and the Praesepe star clusters have been selected to cover a wide range of magnitudes and to have appropriate comparison stars nearby in the eyepiece.

Project 5-1 Spectrum, Luminosity, Parallax and Perspective — cloudy night

This four part project teaches spectral analysis for chemical composition, the inverse square law, and the laws of perspective, parallax , and distance.

Spectrum: The student draws the bright-line spectra of various gases observed with a diffraction grating. The student may then be given a homework assignment to go out and observe the spectrum for various man-made or natural sources to identify their constituent gases by comparing their spectra to the spectra for known gases he/she observed in the classroom.

Luminosity: The student uses a set of grain-of-wheat-bulb lamps set to different intensities. The student places the lamps along a test range at different distances until all appear of equal brightness (to the student's eye). The distances between the lamps are measured and the relative brightnesses of the lamps calculated according to the inverse square law. The emphasis is on visual rather than electronic measurement of brightness to develop the student's intuitive feel for the inverse square law and because the accuracy that can be obtained with the eye is more than sufficient. The instructor sets the lamps to specific brightnesses by performing the project in reverse — place the lamps at prescribed distances and adjust their brightnesses until they appear equal by adjusting a potentiometer in series with the lamp and battery. A more advanced circuit, shown below, will automatically regulate the lamp brightness in compensation for battery drain.

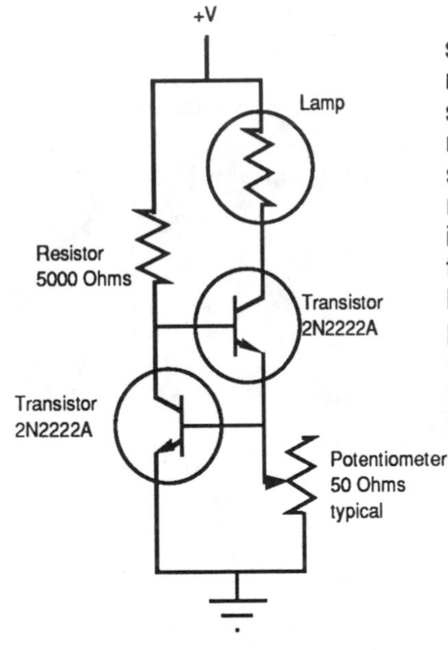

Perspective: The student measures the apparent angular size of a partner's head as a function of distance to establish the relationship between angular size and distance.

Parallax: The student constructs an optical device for measuring the parallactic angle for objects at different distances. Using this device, the student measures the parallax of objects, computes their distances from their parallax, and compares the computed distance to the actual distances.

Project 6-1 Seeing the H-R Diagram — cloudy night

This project teaches the interpretation of magnitude and color of stars, star clusters and galaxies. It is divided into three parts; stars, star clusters, and galaxies.

Stars: The student is introduced to measuring color in stars with color filters. The instructor constructs synthetic stars by making pinholes in a piece of foil mounted in a 35-mm slide mount. Attaching colored plastic filters behind each of the pinholes simulates different color stars which are then projected using a slide projector. As in Project 3-1, it is advisable to use a blue plus a red filter for the synthetic red star to set its intensity comparable to that of the other synthetic stars. A GamColor color filter swatch book can be obtained at nominal cost from:

The Great American Market
826 N. Cole Avenue
Hollywood, CA 90038
(213) 461-0200

Filter transmission functions are provided with the swatch book. Recommended filters for the student are "Daylight Blue" and "Dark Amber." Brightness of the synthetic stars is best adjusted by using a pair of rotating polarizing filters in front of the projector lens rather than varying the projector lamp voltage.

Star Clusters: The student observes color slides of star clusters (generated from published BV photometry) and uses relative magnitudes and colors of their component stars to locate their positions on the H-R diagram. From their location on the H-R diagram, the student can then estimate the relative ages of the star clusters, upper limits for masses of stars on the main sequence, and initial masses of stars in the giant branch. Slides for this project can be obtained from:

Rigel Systems
26850 Basswood
Rancho Palos Verdes, CA 90274
(213) 375-4149

Galaxies: From color slides of several galaxies of different types, the student identifies different populations, compares their locations to that of gas and dust clouds, and estimates the relative ages of the different populations. The best slides for this purpose are those of David Malin, taken with the Anglo-Australian telescope, which can be obtained from:

Hansen Planetarium
15 S. Lake St.,
Salt Lake City, Utah 84111
(801) 534-2104

Project 6-2 Star Search — clear night

This project continues with familiarization with a telescope, including setting up and aligning an equatorial mounting, using the setting circles to find celestial objects, how to observe and what an observation contains, how to interpret colors in terms of temperature and look for any systematic trends in the colors and magnitudes of multiple stars. Most of the questions can be answered from a few observations while a larger number of observations, provides more information. The instructor may supplement observations with magnitude and color data from the Trained Eye Star Atlas or other sources.

Project 7-1 Star Hopping — cloudy night

This project teaches the student to plan an observation session, the tricks of the trade for finding celestial objects as well as how to prove he/she's in the right place and found the right celestial object. The multiple stars chosen give a range of difficulties and interesting color combinations. The non-stellar objects cover a range of types, and difficulties.

Project 7-2 Celestial Sights — clear night

This project hones observing skills by giving the student the opportunity to apply the observing plan developed in Project 6-1 to find different celestial objects and observe their appearance, and interpret their magnitudes and colors in terms of their positions on the H-R diagram. Most of the questions can be answered from a few observations and by including observations from Project 4-3. The instructor may supplement observations with magnitude and color data from The Trained Eye Star Atlas or other sources.

Project A-1 Waiting for the Sun — clear night

Watching sunsets lets the student see how the sun's point of sunset changes in rhythm with the seasons, and how to use these observations to determine the length of a year. The entire course of observations may spread over several months, but require only a few minutes one evening a week.

Project A-2 A Month of Moon — clear night

Watching the moon for a few minutes each day for a month, the student sees how the moon's appearance changes with distance from the sun, how the moon moves from night to night among the constellations, and the length of the synodic and sidereal months.

Project A-3 Wheels Within Wheels — clear night

Because Venus and Mars move so fast from night to night, the student can easily visualize their motions along the ecliptic in his/her mind's eye. Using the unaided eye and telescopic observations of Venus and/or Mars, the student learns how the appearance of the planets, their apparent sizes, their visual magnitudes, and their apparent motions along the celestial sphere relate to their orbital motions and that of the Earth. The student can also use his/her observations to verify the inverse square law. The project is divided into two parts: unaided and telescopic observations. The visual observations require only a few minutes a night for one or two nights a week. Telescopic observations require only a few minutes a week. The instructor may substitute data from the Astronomical Almanac to supplement observations.

Project A-4 Star Gauging — clear and/or cloudy night

By counting stars in different directions, the student can discover the basic shape of the galaxy and estimate our location within it (not accounting for interstellar dust). For a clear-night project, the student makes observations of the celestial sphere. You may wish to combine actual observations with data from a star atlas, constellation star field slides, or a planetarium to complete coverage of the celestial sphere. As a cloudy-night exercise, all data could be obtained from these last three sources.

Project A-5 Intrinsic Variable Stars — clear night

Telescopically observe variable stars change in luminosity and relate these changes to their positions on the Hertzsprung-Russell and period-luminosity diagrams. Variables have been selected to encompass a range of types (Population I and II Cepheids, RR Lyrae, RV Tauri, Semi Regular) and periods of variation (hours to months — from one class night to a semester), have easily seen variations (0.7+ magnitudes), and comparison stars within ± 1.5 degrees (easy to see in a telescope). The American Association of Variable Star Observers (AAVSO) publishes predictions of times of maxima for variable starsthat can assist the instructor in selecting stars. The address for the AAVSO is:

AAVSO
25 Birch Street
Cambridge, MA 02138
(617) 354-0484

Project A-6 Eclipsing Variable Stars — clear night

Telescopically observe eclipsing variables to see stars orbiting about each other. Eclipsing variables have been selected to with periods of variation from hours to a few days (from one to several class nights), have easily seen variation (0.7+ magnitudes), and comparison stars within ± 1.5 degrees. The American Association of Variable Star Observers (AAVSO) annually publishes predictions of times of minima for eclipsing binaries that can assist the instructor in selecting stars.

Acknowledgements

No book can be written without the help of others and this book is no exception. I am most deeply in debt to my wife Linda for her loving assistance and encouragement. Thank you Linda for your long hours of reading, proofing, editing. Thank you for your critique and suggestions, for your collaboration. I must also thank Bob Shadduck and Paula Evron for assistance in generating the charts for variable starts as well as Dave Capka and Hans Mahr for helping put together the charts for the star clusters M67, M44 and M45. I also deeply appreciate the assistance and perseverence of my editors, Jeff Holtmeier and Kate Pachuta . Finally, thanks to the El Camino College students who tried, tested and proved the material that has gone into the text and projects.

Contents

	Instructor's Preface and Guide		*v*
	Invitation to Discovery		*1*
Chapter 1	*The Pattern of Stars*		*3*
	Project 1-1	*Charts, Stars and Magnitudes*	*15*
	Project 1-2	*Mapping the Heavens*	*21*
Chapter 2	*The Mortal's Point of View*		*25*
	Project 2-1	*The Night Sky*	*39*
	Project 2-2	*Diurnal Motions*	*43*
Chapter 3	*Extending the Eye's Reach*		*49*
	Project 3-1	*Basic Astronomical Optics*	*59*
Chapter 4	*Establishing a Firm Foundation*		*77*
	Project 4-1	*Telescope Field of View and Light Gathering Power*	*87*
	Project 4-2	*Visual Photometry of Stars*	*95*
Chapter 5	*Taking the Measure of a Star*		*105*
	Project 5-1	*Spectrum, Luminosity, Parallax and Perspective*	*117*
Chapter 6	*Reading the Message in Starlight*		*129*
	Project 6-1	*Seeing the H-R Diagram*	*143*
	Project 6-2	*Star Search*	*153*
Chapter 7	*Observing*		*163*
	Project 7-1	*Star Hopping*	*169*
	Project 7-2	*Celestial Sights*	*181*

Appendix A Additional Projects 191
Project A-1 Waiting for the Sun 193
Project A-2 A Month of Moon 195
Project A-3 Wheels Within Wheels 197
Project A-4 Star Gauging 205
Project A-5 Intrinsic Variable Stars 211
Project A-6 Eclipsing Variable Stars 237

Appendix B Amateur Astronomy Organizations 251

Appendix C Trained Eye Star Atlas 261

Invitation to Discovery

Go out on a clear evening and watch simplicity masquerade as complexity — in the confusion of stars scattered across the hemisphere of night. Resist the urge to flip on the porch light — make an astronomical discovery instead. Pick out a pattern of a few bright stars. Blink and the pattern is still there. Come back the next evening and you'll find the same pattern waiting for you. You've discovered the *fixed* (unchanging) nature of the patterns of stars on the *celestial sphere* (sphere of stars). More importantly, you've begun the process of discovering order in chaos — the simplicity that underlies the complexity of the natural world. All it takes is a trained eye.

The trained eye is really the mind behind the eye, with its ability to discover the laws of physics and use them to understand what the eyes see. It is a natural talent, as easy as catching a baseball — which you could not do without an "intuitive" knowledge of basic laws of physics. Just think about what is involved in catching a baseball. The mind coordinates both eyes to track the baseball and uses their input, along with other sensory clues, to calculate the baseball's position, velocity, and acceleration. The mind anticipates the baseball's mass and inertia, uses the law of gravitation to predict it's future trajectory, and commands a hand to move to the right location in space-time to catch it.

We don't carry Newton's formal mathematical description for the physical laws of motion and gravitation around in our heads, we understand them intuitively. Even Aristotle, whose "logical" but erroneous laws of motion and gravitation dominated the world for the 2000 years preceeding Newton, intuitively knew the correct laws. In his writings, he hailed logic. In everyday experience, he, like the rest of us, trusted intuition.

If you can catch a baseball, you can understand astronomical observing in the same way — *intuitively.* This book helps achieve that goal by teaching, not just describing, by presenting exercises that develop intuitive skills, by providing a framework around which understanding can grow. All you need is a trained eye.

Where We Go from Here

This book is a training program in which you learn by doing. Hands-on projects illuminate the topics covered in each chapter. Chapter 1 begins by reintroducing the celestial sphere as an old friend and continues with how to capture and preserve knowledge about the celestial sphere. Chapter 2 builds upon Chapter 1 by bringing the revolving, rotating Earth back into the celestial sphere. Once you've absorbed the structure and apparent motions of the celestial sphere, it's time to turn to observing. Chapter 3 provides a tutorial on basic telescope optics — the parts of the telescope you look through. However it takes more than just something to look through, you need a rock to rest it on — hence Chapter 4 on telescope mountings. When you've finished Chapter 4, you know the celestial sphere and the telescope but not technique — the skills that separate the apprentice from the master. Chapter 5 and 6 reveal the physics behind the observation — show how to read the message in starlight. Finally, Chapter 7 brings all your skills together and hones your observing technique until it is second nature.

Chapter 1
The Patterns of Stars

Nobody goes out, looks up, and "bingo" identifies a star. You look for a particular pattern of stars on the celestial sphere, and then a particular star in the pattern. The key word is *pattern* ; the key skill is *pattern recognition* — the ability to find (or create) structure in randomness, as in Figure 1-1.

At first glance, Figure 1-1 looks like a random jumble of light and dark blotches, but there is a pattern hidden in the jumble. The word "cow" will bring an image to mind, a mental image that matches the pattern in Figure 1-1. This exercise in perception illustrates what you need to learn the patterns of stars on the celestial sphere: common (recognized by everyone) patterns with common names. The common patterns of stars are called *constellations,* with a total of 88 constellations mapping the entire *celestial sphere.* Table 1-1 lists the constellation names and their abbreviations.

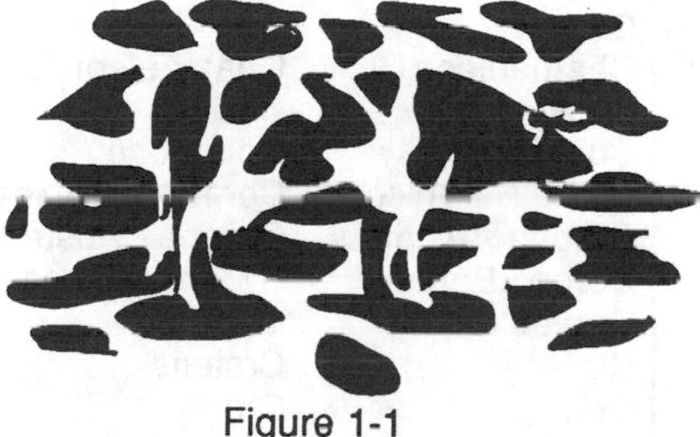

Figure 1-1

The common names for most of the constellations are pretty uncommon. Most of the names for the ancient constellations originated so long ago that we don't know who invented the names or what patterns they saw on the celestial sphere that inspired the names. All we have are the names and their stories; stories that may tell of great deeds and great heroes, gods and mortals, but not what pattern to look for. The modern constellations, added since the 1700's, mostly to fill in constellations for the southern stars, lack the pedigree of the ancient constellations and bear even less resemblance to their namesakes. Figure 1-2 illustrates the problem of recognizing constellations from their names and illuminates the solution.

The stars in Figure 1-2 compose the ancient constellation *Ursa Major* — the "*Big Bear.*" The problem is finding a bear of any size in the pattern of the stars forming *Ursa Major.* What does pop out? The *pattern* of the brighter stars called the "*Big Dipper.*" That's the pattern you see in Figure 1-2 ; that's the pattern you'll see at night. Many constellations have such "*nickname*" patterns of their bright stars : Ursa Major the "*Big Dipper,*" *Ursa Minor* the "*Little Dipper,*" the "*Teapot* " of *Sagittarius*, and *Cygnus* the "*Northern Cross.*" These bright

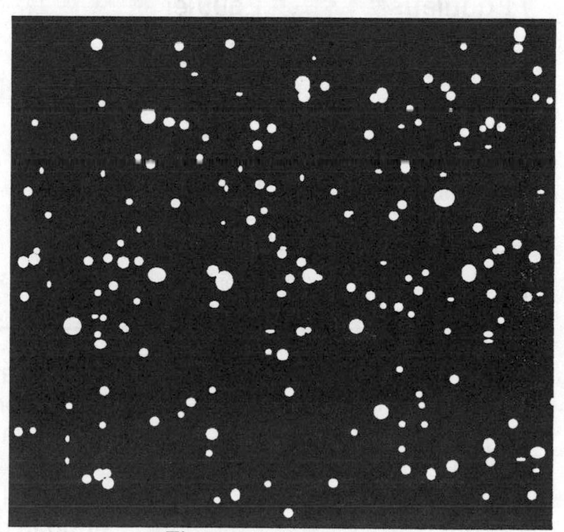

Figure 1-2

Table 1-1: The Constellations

Constellation	Possessive	Abbrv.	Constellation	Possessive	Abbrv.
Andromeda	Andromedae	And	Lacerta	Lacertae	Lac
Antlia	Antliae	Ant	Leo	Leonis	Leo
Apus	Apodis	Aps	Leo Minor	Leonis Minoris	LMi
Aquarius	Aquarii	Aqr	Lepus	Leporis	Lep
Aquila	Aquilae	Aql	Libra	Librae	Lib
Ara	Arae	Ara	Lupus	Lupi	Lup
Aries	Arietis	Ari	Lynx	Lyncis	Lyn
Auriga	Aurigae	Aur	Lyra	Lyrae	Lyr
Boötes	Boötis	Boo	Mensa	Mensae	Men
Caelum	Caeli	Cae	Microscopium	Microscopii	Mic
Camelopardalis	Camelopardalis	Cam	Monoceros	Monocerotis	Mon
Cancer	Cancri	Cnc	Musca	Muscae	Mus
Canes Venatici	Canum Venaticorum	CVn	Norma	Normae	Nor
Canis Major	Canis Majoris	CMa	Octans	Octantis	Oct
Canis Minor	Canis Minoris	CMi	Ophiuchus	Ophiuchi	Oph
Capricornus	Capricorni	Cap	Orion	Orionis	Ori
Carina	Carinae	Car	Pavo	Pavonis	Pav
Cassiopeia	Cassiopeiae	Cas	Pegasus	Pegasi	Peg
Centaurus	Centauri	Cen	Perseus	Persei	Per
Cepheus	Cephei	Cep	Phoenix	Phoenicis	Phe
Cetus	Ceti	Cet	Pictor	Pictoris	Pic
Chamaeleon	Chamaeleontis	Cha	Pisces	Piscium	Psc
Circinus	Circini	Cir	Piscis Austrinus	Piscis Austrini	PsA
Columba	Columbae	Col	Puppis	Puppis	Pup
Coma Berenices	Comae Berenices	Com	Pyxis	Pyxidis	Pyx
Corona Austrinus	Coronae Austrini	CrA	Reticulum	Reticuli	Ret
Corona Borealis	Coronae Borealis	CrB	Sagitta	Sagittae	Sge
Corvus	Corvi	Crv	Sagittarius	Sagittarii	Sgr
Crater	Crateris	Crt	Scorpius	Scorpii	Sco
Crux	Crucis	Cru	Sculptor	Sculptoris	Scl
Cygnus	Cygni	Cyg	Scutum	Scuti	Sct
Delphinus	Delphini	Del	Serpens	Serpentis	Ser
Dorado	Doradûs	Dor	Sextans	Sextantis	Sex
Draco	Draconis	Dra	Taurus	Tauri	Tau
Equuleus	Equulei	Equ	Telescopium	Telescopii	Tel
Eridanus	Eridani	Eri	Triangulum	Trianguli	Tri
Fornax	Fornacis	For	Triangulum Australe	Trianguli Australe	TrA
Gemini	Geminorum	Gem	Tucana	Tucanae	Tuc
Grus	Gruis	Gru	Ursa Major	Ursae Majoris	UMa
Hercules	Herculis	Her	Ursa Minor	Ursae Minoris	UMi
Horologium	Horologii	Hor	Vela	Velorum	Vel
Hydra	Hydrae	Hya	Virgo	Virginis	Vir
Hydrus	Hydri	Hyi	Volans	Volantis	Vol
Indus	Indi	Ind	Vulpecula	Vulpeculae	Vul

constellations with their recognizable nickname patterns provide the framework for learning the celestial sphere, while faint stars and faint constellations fill out the framework.

Still, you can look as long as you want; you won't see any signs hung in the sky labeled Ursa Major, nor glowing lines interconnecting the stars in the Big Dipper. We, not nature, impose this framework of constellations upon the celestial sphere: it is our creation — our model — that reveals structure in nature. In nature, structure is hidden; in our model, we have made it obvious. Still, there are many different ways to structure the celestial sphere, so why constellations? Because the only reliable storage media available that could pass this knowledge over the centuries were the minds of longvanished storytellers. The framework of constellations is their invention: a model optimized for human memory.

Written versions of the constellation stories now exist but, unlike the ancient storytellers, printed words cannot point out the constellations they describe. Without a human guide, you need a physical model of the celestial sphere to show the connection between the constellations and the stars.

Modeling the Celestial Sphere

The medium we choose for the physical model is paper, the technique we choose is mapping. Think for a moment about how to make a map of the celestial sphere. First, what surface should the celestial sphere be drawn on? The best map is that which represents the celestial sphere with the least distortion. In addition, there is the choice of symbols to represent stars, other celestial objects and the structure you wish to reveal. The best choice for symbols is the obvious choice — pick symbols that are easier to read than the stars.

The best surface to map the celestial sphere onto is another sphere. Spherical maps, however, have serious problems. To show detail, they must be large; for portability, small.

Box 1 1: Perception and Reality

As you begin, there is one particular trap to avoid; do not confuse models with reality. For example, the celestial *sphere* is a model of reality, not reality itself. We are not sitting in the middle of some giant beach ball with spots of luminescent paint splattered across the inside. As a model, however, thinking of the stars as attached to the inside of a celestial sphere lets us apply our intuitive knowledge about spheres to what we see (if it looks like a sphere and acts like a sphere ...). The use of such models is fine so long as the models explain *everything* we observe. When we observe something that a model can't explain, then it is time to change the model, not reality.

How can this paradox be resolved? You could make a big detailed globe and flatten it into a sheet. Although this solves the detail and portability problem, it introduces a new dilemma — distortion. Spheres do not flatten nicely, as you can prove to yourself by deflating a basketball.

At the other extreme, you can chop the model celestial sphere into a myriad of tiny pieces. Tiny pieces do not distort upon squashing since they're already pretty much flat. You have preserved portability and detail with no distortion, but now have the jig-saw puzzle problem of figuring out how all the pieces fit back together.

Celestial sphere maps using both of these techniques are available on the market. Planispheres follow the first approach, jig-saw puzzle Star

Atlases the second, and both suffer accordingly. *The Trained Eye Star Atlas* (accompanying this volume) takes a middle road to success. The Trained Eye Star Atlas has maps sized to balance detail and portability. The entire celestial sphere is divided into eight maps, each showing the connectivity between constellations with minimal distortion. To make these maps, slice the celestial sphere just as you would peel an orange (Figure 1-3).

There are two types of peels for the orange, so two types of maps in The Trained Eye Star Atlas: two *polar maps* for the stars near the *north* and *south celestial poles* and six *equatorial maps* girdling the middle of the celestial sphere (more about poles and equators later). Now join these maps back into a model celestial sphere. Mentally cut the maps out (Figure 1-4). Tape each equatorial map to the left side of the one preceeding it until they circle around you and the last equatorial map butts against the first. Tape the north polar map over the top of the equatorial map ring, making sure that the constellations line up properly. Tape the south polar map underneath and you're done.

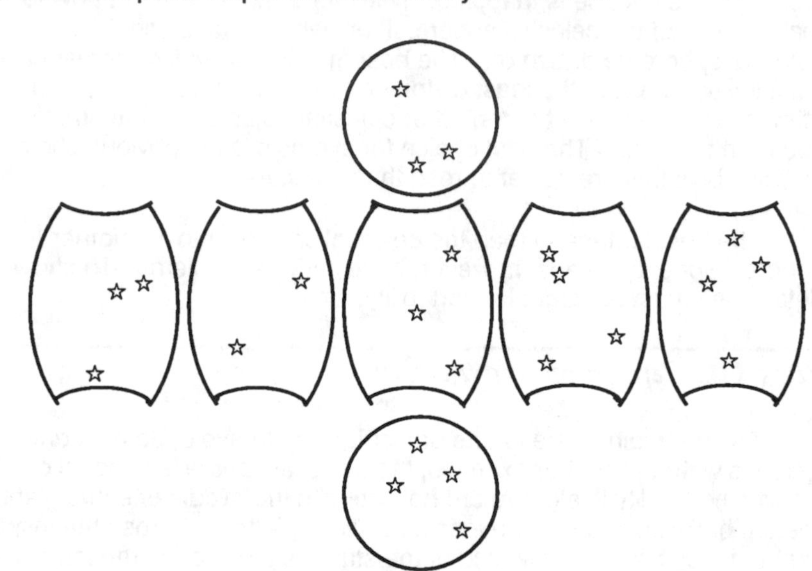

Figure 1-3

So much for the form for the maps in The Trained Eye Star Atlas. Now what about the symbols on the maps? Look at the north polar map and you'll see the names for constellations printed right on the map. Ursa Major is there — so is Ursa Minor. Dashed lines indicate the borders between constellations — where one constellation ends and another begins. The nickname patterns for the constellations are not drawn in yet for any of the constellations. You get to do that in Box 1-3. Stars appear as circles of different sizes on the maps. However, we require more than just circles indicating the existence of stars; we need to tell them apart.

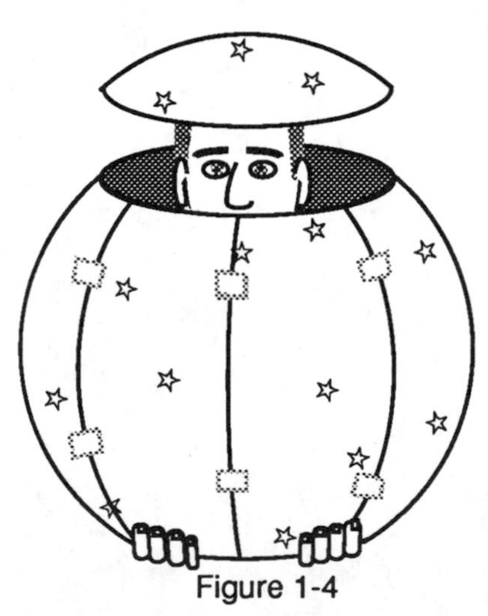

Figure 1-4

The Stars in Patterns

The Trained Eye Star Atlas presents two additional pieces of information for each star: *name* and *apparent visual magnitude* . Most of the *common names* we use are Greek, Roman or Arabic in origin and are often very melodious; *Mira* "*the wonderful,*" for example. Still, there are thousands of stars that you can see with the unaided eye and so potentially thousands of common names (not to mention hundreds of misspellings). Also, what do you do about assigning stars to constellations? For a long time, astronomers used a convenient trick: "Not enough bright stars to make a constellation look like something recognizable? Just borrow a few from the neighbors." As long as star names were independent of constellations then the stars themselves were independent of the framework of constellations.

The astronomer *Johann Bayer* devised a solution in 1603 — make the constellation name a part of a star's name and fix the stars to the framework of constellations. Bayer assigned Greek alphabet letters to the stars he could see in each constellation according to brightness (there are exceptions — find them and figure out what convention he used). As an illustration of Bayer's naming convention, *Vega*, the brightest star in the constellation of *Lyra* is α *Lyrae*, the second brightest star in Lyra is β *Lyrae*, and so on. Unfortunately, most constellations have more stars than there are Greek letters. So when Bayer ran out of Greek letters, he switched to lower case Roman letters (a,b,c ...). When he ran out of those, he switched to upper case Roman letters (A, B, C...). Fortunately, he ran out of stars before he ran out of letters.

In 1725 *John Flamsteed*, another astronomer, devised a similar scheme that avoided the limitations of alphabets by numbering the stars in each constellation from *west* to *east*. For example, the Flamsteed names for *Vega* and β *Lyrae* are 3 *Lyrae* and 10 *Lyrae*, respectively. Although he never ran out of numbers, Flamsteed's names do not carry the brightness information for stars that the Bayer names do, and precession (an effect, due to the gravity of the Sun and the Moon on the Earth that causes the north pole to wobble like a top over 26,000 years) has somewhat scrambled the west-to-east ordering over the last 263 years.

Astronomers have since decided to use a hybrid of the Bayer and Flamsteed naming conventions: Bayer names for stars that have Greek letters, and Flamsteed numbers for the rest. That's what you see next to each star in The Trained Eye Star Atlas. Still, these official names are cold — *Algol* "*the demon star*" is more intriguing than β *Persei*. So look for the common names for the brighter stars adjacent to the Bayer-Flamsteed names in The Trained Eye Star Atlas.

One last note on names. Look what happens to the word Lyra when we incorporate it into Vega's Bayer-Flamsteed name of α Lyrae. The ending changes to indicate possession; the star α belongs to the constellation Lyra. Similarly, Perseus becomes possessive in β Persei. Table 1-1 lists the possessive forms for the constellation names.

Now on to the other parameter we need for stars, *apparent visual magnitude*. Such a mysterious phrase. Well, actually only one of the three words is unfamiliar. *Apparent* means what it says, how does it "*appear*" from where you're standing. If you're standing upon the Earth, it will appear quite differently than if you're standing on a planet going round a distant star. "*Visual*" is equally familiar— it means "*as seen with the human eye* ." So we are left with the mystery word "*magnitude*". You probably already have some feeling for magnitude's usage, but astronomers use it when talking about *brightness* of stars. Why not just use apparent visual brightness? The eye is not a simple window through which the mind peers out at the universe. The eye is an extension of the brain, a pre-processor which reduces the flood of visual stimulus to a rivulet of sensation that the mind can accommodate. During this preprocessing, the eye converts light from brightness into magnitude. It is magnitude which the mind perceives (see Box 1-2).

Hipparchus, an ancient Greek astronomer, devised the original *apparent visual magnitude* scale for stars (back around 120 BC) as an aid in telling stars apart. He assigned *first magnitude* to the *brightest* stars he could see, *sixth magnitude* to the *dimmest*, and magnitudes in between to the stars in between. Astronomers have since revised and refined Hipparchus' original apparent visual magnitude scale and codified the relationship between magnitude and brightness. Still, Hipparchus' apparent visual magnitude scale lives on and has been the seed for other magnitude scales (e.g., *bolometric* and *absolute magnitudes*). Because there are other magnitude scales, we have to be careful in the use of the word magnitude. For now, the word magnitude is used in the context of apparent visual magnitude.

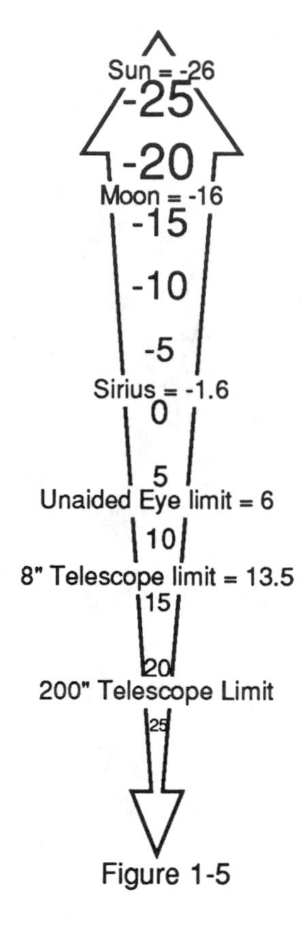

Figure 1-5

Let's look at the magnitude scale in detail. Reiterating Hipparchus' scheme, the smaller the magnitude, the bigger the brightness of the star. Smaller is bigger? How odd! Not really. Think back to your high school days — what was the brightest student in your class ranked? And the second brightest? And the third brightest? That's the way to think about magnitudes. Figure 1-5 illuminates the magnitude scale and its extension beyond Hipparchus' original 1-to-6 magnitude range.

The circles, representing stars on the maps in The Trained Eye Star Atlas, encode magnitudes in two ways. First, the "*size*" of a circle indicates a star's magnitude. If you compare the visual magnitude of two stars;

The star with the *bigger* circle has the *brighter* magnitude.
The star with the *smaller* circle has the *dimmer* magnitude.

Using circle sizes to represent magnitudes gives the star maps the correct look while the fillings of the circles change every half-magnitude to make reading an individual star's magnitude easier.

Assigning apparent visual magnitudes and names to stars completes the mapping of the celestial sphere that started with assigning stars to the constellations. There are other ways of mapping the celestial sphere (as you'll see in Chapter 1-4), but none as intuitive. Remember, the framework of constellations originated long before writing. It has passed down through generations by the only method possible — storytelling. Storytelling is a very human method of encoding knowledge in a form that human minds can learn, recall and transmit without error.

Box 1-2: Perception and the Magnitude Scale

The capabilities of the eye far surpass those of the best camera. Your eye is so sensitive it can sense a single photon (packet of light energy) or accomodate the glare of the noon-day Sun, some 100 billion times as intense. Logically, this is surprising since, given that a single photon generates one nano (one billionth) amp of current, shouldn't the brightest glare of the noon-day Sun send 100 amps (100 billion nano amps) up your optic nerves into your brain? Logic must be wrong.

You can avoid this sensory paradox by avoiding *linear* logic; your eye doesn't have to generate 100 billion times as much *sensation* (the information that it transmits along the optic nerve to the brain) for a 100 billion time increase in *stimulus* (the light coming into your eye). Evolution discovered this trick a long time ago. Your eye does more than just sense light, it converts linear changes in stimulus into *logarithmic* changes of sensation. For example, when you watch a light become 10,000 *times brighter*, your eyes only produce a 10 *magnitude difference* in sensation — multiplication converted into addition. The equations that describe this relationship are:

$$\text{magnitude difference} = -2.5 * \log_{10}(\text{brightness ratio})$$

and going the other way:

$$\text{brightness ratio} = 10^{\left(\frac{\text{magnitude difference}}{-2.5}\right)}$$

The -2.5 tweaks the equations to match Hipparchus' original visual magnitude scale with smaller magnitude being brighter and larger magnitude being fainter.

Another way to visualize the relationship between brightness ratio and magnitude difference is to convert the equations to a table (Table 1-2). Table 1-2 is the equation — for the values listed. A *negative* magnitude difference implies comparing a *brighter* star to a *fainter* star, a *plus* magnitude difference implies comparing a *fainter* star to a *brighter* star. Note that the choice of brighter or fainter implies two ways of comparing stars that result in the same answer. The order of comparison, revealed by the sign of the magnitude difference, is important in using the table correctly.

For example, star A is magnitude 10 while star B is magnitude 15. What is the brightness ratio of the two stars? Start by figuring which order you're going to compare them: star A to star B or star B to star A.

Comparing star A to B, you would subtract 15 from 10 to get a magnitude difference of -5. The brightess ratio in the table that corresponds to a magnitude difference of -5 is 100; star A is 100 times brighter than star B.

Comparing star B to star A, you would subtract 10 from 15 to get a magnitude difference of +5. The brightness ratio in the table that corresponds to a magnitude difference of +5 is 1/100th; star B is 1/100th as bright (100 times fainter) than star A.

Another example. Star C is 1000 times brighter than star D of magnitude 20. What is the visual magnitude of star C? Going to the table, a brightness ratio of 1000 corresponds to a magnitude difference of -7.5 magnitudes. Add the -7.5 to star D's visual magnitude of 20 and the visual magnitude of star C is 12.5.

Table 1-2

Magnitude Difference	Brightness Ratio
+16.5	1/4,000,000
+15	1/1,000,000
+13.5	1/400,000
+12.5	1/100,000
11.5	1/40,000
+10	1/10,000
+9	1/4,000
+7.5	1/1,000
+6.5	1/400
+5	1/100
+4	1/40
+2.5	1/10
+1.5	1/4
+1	1/2.5
+0.5	1/1.6
0	1
-.5	1.6
-1	2.5
-1.5	4
-2.5	10
-4	40
-5	100
-6.5	400
-7.5	1,000
-9	4,000
-10	10,000
-11.5	40,000
-12.5	100,000
-14	400,000
-15	1,000,000
-16.5	4,000,000

Beyond the Framework of Constellations

With the invention of writing, "*paper memory*" supplemented human memory, and the legends of the constellations passed from storytelling to written record. However, paper memory did not just supplement human memory — as a new technology, it presented new possibilities for encoding knowledge. It was now possible to map the celestial sphere in new, more detailed ways, relying on paper memory, not human memory, to preserve the knowledge. Still, the mere existence of a new technology does not reveal its usefulness. Astronomers devised new mapping techniques based on paper memory because of the failure of the old technique based on human memory.

The framework of constellations failed at specifying the locations for stars and other celestial objects with the accuracy that civilizations needed to measure time, regulate the calendar, and predict the future. In many ancient civilizations, the astronomers were priests and the planets gods. The stars, fixed to the celestial sphere, provided the backdrop against which the gods/planets danced their passion play. The priest/astronomers discovered that they could predict the movements of the gods/planets in their dance, i.e., predict their future. The predictions weren't perfect at first, but surely that would come with exacting observation and diligent study. The priest/astronomers believed that once they could accurately predict the future for the gods/planets, then predicting the future for mere mortals would be trivial.

This desire to know the future drove the priest/astronomers to find ways of more accurately mapping the celestial sphere than the framework of constellations. The particular priest/astronomers we have to thank were the Babylonians, for it is they who cursed us with 60 seconds to the minute, 60 minutes to the hour, and 24 hours to the day. Why didn't they just count by tens like common folk? Probably because they weren't commoners. These priest/astronomers owed their prestige and livelihoods to their knowledge of the divine heavens. Knowledge that was, and remains, power — not to be given away freely. What better way to protect sacred knowledge from profane minds than a celestial counting ritual mandated by the gods.

Coordinating the Stars

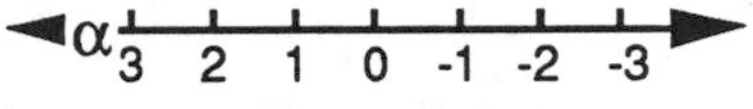

α 3 2 1 0 -1 -2 -3

Figure 1-6

The secret to the celestial sphere that the Babylonian priest/astronomers discovered was powerful in its simplicity — the position of any star could be specified exactly by using just *two coordinates*. What are *coordinates*? A few commonplace examples of *coordinate systems* give the best definition: chessboards (queen's knight to queen's bishop 4), crossword puzzles (28 across) and the ubiquitous bingo (B 14). A *coordinate* is just some measure of distance along some *axis* (line) in some direction from some *origin* (starting point). The simplest *coordinate system* imaginable is the one-dimensional "*number axis*" (Figure 1-6). On this *number axis*, the *coordinate* measures the number of units something is left (plus) or right (minus) of the *origin* at zero along the *number axis*. For example, the coordinate of α is 3.5, which tells you α is located 3.5 units to the left of zero.

A two-dimensional surface (like a sheet of graph paper) requires two number axes (axes is the plural of axis) oriented perpendicular to each other (Figure 1-7). The *coordinates* for any point on the sheet is just an ordered pair of numbers — the horizontal (left/right) coordinate followed by the vertical (up/down) coordinate. On the two-dimensional surface the coordinates for α are (+3.7, -2.5) which means go left 3.7 and down 2.5.

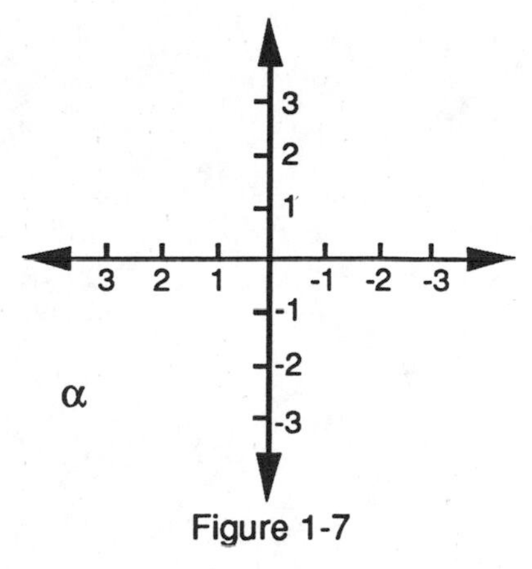

Figure 1-7

What works for flat surfaces has equivalents for curved surfaces, as the Babylonian priest/astronomers discovered. Equivalent does not mean identical — you don't just take a sphere and crumple a piece of graph paper around it. The best way to understand the creation of the *celestial coordinate system* is to look at the celestial sphere from the god's-eye perspective — from the outside (Figure 1-8).

The first thing to do to create a coordinate system on the celestial sphere is to pick a point. So pick a point and call it the *north celestial pole* — as good a name as you'll see in Chapter 2. Pick another point directly opposite the north celestial pole on the celestial sphere and call it the *south celestial pole*. Then draw a line on the celestial sphere that stays exactly halfway between the north and south celestial poles. This line will circle all the way around the *celestial* sphere and divide it into two *equal* halves, so call it the *celestial equator*. Pick a point on the celestial equator and call it the *origin* by assigning it the coordinate value zero and the name *"first point of Aires"* (the origin of this point and its name will be revealed in Chapter 2). Then tick another 23 evenly spaced marks around the celestial equator and number them 1 to 23. The celestial equator now marks *right ascension* one of the two coordinates for the celestial coordinate system. *Right ascension* measures the distance around the celestial equator in units of *hours* (*h*), *minutes* (*m*) and *seconds* (*s*) (thank you, Babylonians), with 24 *hours* equaling a full circle around the celestial equator. Why measure a circle in units of time? The question is backwards — time measures the angular distance around the celestial equator to the Sun, an angle measured in hours, minutes and seconds.

You still need a second coordinate to measure distance *up* and *down* from the celestial equator. *Up* and *down*? They only make sense when used to describe directions from our mortal viewpoint inside the celestial sphere, not on it. To wit: *up* means from the ground towards a point overhead, and *down* means from a point overhead to the ground. We need new words to describe directions *on* the celestial sphere, so let's "invent" the terms *north, south, east* and *west* (Figure 1-9). *East* is the direction of increasing *right ascension*, and *west* is the opposite direction to *east* (in the direction of decreasing *right ascension*). *North* is as shown, and *south* is the opposite direction to *north*.

Now generate the second coordinate by drawing lines due north and south away from the celestial equator at each hour of right ascension. A funny thing happens to the lines drawn to the north — they converge to a point. So do the lines drawn to the south. These are the two points with special names, the *north celestial pole* and the *south celestial pole*.

The second celestial coordinate, *declination*, measures the distance of a star north or south of the celestial equator in *degrees* (°), *arc-minutes* (') and *arc-seconds* (") (thanks again, Babylonians), from 0° at the celestial equator to +90 ° at the *north celestial pole* or -90 ° at the *south celestial pole*. Notice you can't have a declination greater than +90° or less than - 90 °. Once you cross over the

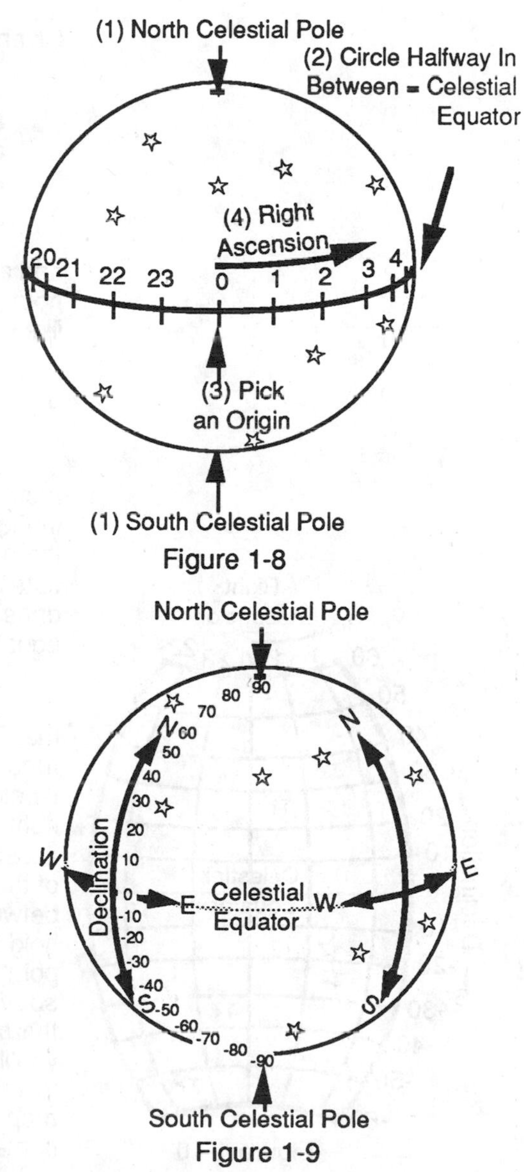

Figure 1-8

Figure 1-9

north pole you're on the other side and heading south, so declination decreases.

Notice that the key to defining this celestial coordinate system is picking the point for the north celestial pole. It's choice seems arbitrary only because the chooser hasn't been properly introduced (i.e., the *Earth*, Chapter 2). Also, don't confuse arc-second (") and arc-minute (') with second of time (s) and minute of time (m);

An arc-second (")	equals	1/60th of an arc-minute (') ,
an arc-minute (')	equals	1/60th of a degree (°), and
a degree (°)	equals	1/360th of a full circle.

In contrast,

A second (s)	equals	1/60th of a minute (m),
a minute (m)	equals	1/60th of an hour (h), and
an hour (h)	equals	1/24th of a full circle.

Given this difference in encompassing a full circle $\left(\frac{24}{360}\right)$,

an arc-second (")	equals	1/15th of a second (s),
an arc-minute (')	equals	1/15th of a minute (m) , and
a degree (°)	equals	1/15th of an hour (h).

To keep this distinction straight, just use the terms arc-second (") and arc-minute (') with degree (°) and second (s) and minute (m) with hour (h). Now its time to see what the celestial coordinate system looks like from the mortal's-eye view — from back inside the celestial sphere.

The Insider's View

A sharp knife cuts open the celestial sphere and reveals the mortal's-eye viewpoint. The choice of cuts is familiar, just like the orange in Figure 1-5. This time as you unfold and flatten the peels they become the maps in The Trained Eye Star Atlas (Figure 1-10,11). The choice of cuts resulted in two types of peels, hence two types of maps. Two round ones — the north and south polar maps, and six elongated ones — the equatorial maps.

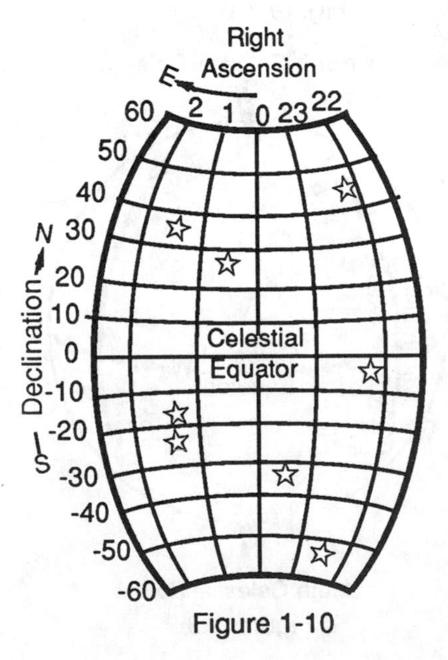

Figure 1-10

Right ascension and declination are encoded on the two types of maps in different ways. On the equatorial maps, the celestial equator runs across the middle of the maps and the hours of right ascension are labeled at the top and bottom. The 11 tic marks between each hour of right ascension correspond to each 5 minutes of each hour of right ascension. Each 10 ° of declination is printed on the left and right edges of the equatorial maps for the range of -60 to +60 degrees. Each tic mark between the tens of declination corresponds to one degree. As you hold a equatorial map and look at it, *up* is *north* and *down* is *south* . A point of caution: *up* and *down* on the maps correspond to *north* and *south* on the celestial sphere only because people always hold maps with the *north* side *up.* In the sky, they are definitely not synonymous (as you'll learn in Chapter 2). Right ascension on the equatorial maps increases to the *left*, so *left* is *east* and *right* is *west.* Contrast this to a map of the Earth where east and west are right and left. The difference comes from looking at the Earth from the outside, not from the inside as we do the celestial sphere. Since *left* is *east* (on the maps), each equatorial star map goes to the left of its predecessor (Figure 1-10) and overlaps it by 1 hour of right ascension for continuity from one map to the next .

The north and south polar maps also encode right ascension and declination but in a strikingly different way (Figure 1-11). On these maps, right ascension runs around the edge — *clockwise* on the north polar map and *counterclockwise* on the south polar map (an important distinction to keep in mind). Thus, *eastward* is *clockwise* on the north polar map and *counterclockwise* on the south polar map. Declination varies radially inward on the maps from ±50 at the edges to ±90 at the centers. So *northward* is *inward* and *southward* is *outward* on the north polar map, while on the south polar map *northward* is *outward* and *southward* is *inward* (the other way around). The polar maps also overlap the equatorial maps by 10 degrees of declination for continuity.

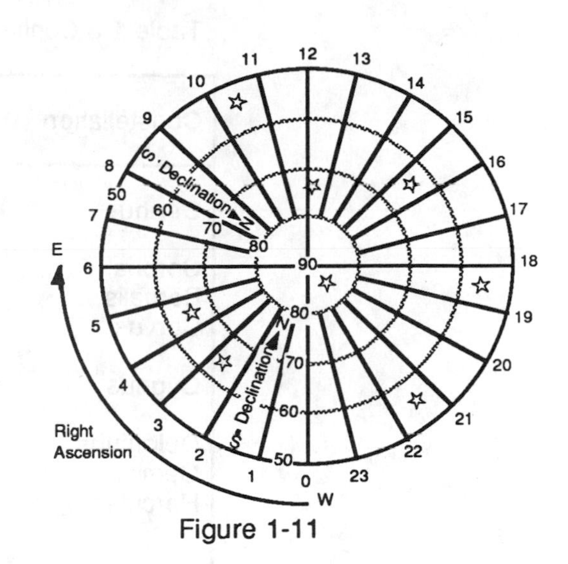

Figure 1-11

Box 1-3: Connect the Dots

Looking at the maps in the The Trained Eye Star Atlas, you are probably overwhelmed by all the dots. The best way to recognize constellations is to connect the dots in the constellations so that they look like something. In the Table 1-3 suggests basic constellation patterns. Find the stars indicated by the Greek letters in each constellation and connect them in the order listed.

Sometimes the result will be a closed shape (the starting point and the ending point are the same star), sometimes an unclosed shape, and sometimes a combination of either or both. Use a red pencil and a ruler to connect the dots. You can find the map for each constellation by consulting The Trained Eye Star Atlas.

Table 1-3 : Constellation Connect The Dots

Constellation	Abrv.	Pattern	Connect the Dots
Andromeda	And	Big V	51–μ–π–α–δ–β–γ
Aquila	Aql	Loop	α–γ–μ–δ–ν–η–β–α
Aries	Ari	Triangle	α–β–γ–α
Auriga	Aur	Helmet	α–ε–ζ–η–ι–θ–β–α
Bootes	Boo	Kite	α–ρ–γ–β–δ–ε–α
			α–η–τ–ν–η
Canis Major	CMa	Big Dog	α–ι–γ–θ
			α–β
			α–π–δ–η
			δ–ε
Canis Minor	CMi	Dash	α–β
Cassiopeia	Cas	Big W	β–α–γ–δ–ε

Table 1-3 Continued: Constellation Connect The Dots

Constellation	Abrv.	Pattern	Connect the Dots
Cephus	Cep	House	α–β–γ–ι–ζ–α β–ι
Corona Borealis	CrB	Diadem	θ–β–α–γ–δ–ε
Corvus	Crv	Crow	α–ε–γ–δ–β–ε δ–η
Cygnus	Cyg	Cross	α–γ–η–β ζ–ε–γ–δ–κ
Delphinus	Del	Dolphin	ε–β–α–γ–δ–β
Gemini	Gem	Twins	α–τ–ε–μ–ν–γ–ζ–δ–ν–β
Hercules	Her	Keystone	α–δ–ε–π–ι β–ζ–η–σ–τ η–π ζ–ε
Leo	Leo	Sphinx	α–η–γ–δ–β–θ–α–ο γ–ζ–μ–ε
Lyra	Lyr	Parallelogram	α–ζ–δ–γ–β–ζ
Orion	Ori	Hourglass	α–γ–δ–β–κ–ζ–α ζ–ε–δ
Pegasus	Peg	Great Square	α–β–α And-γ–α
Persus	Per	Conehead	β–κ–ι–τ–η–γ–α–δ–ν–ε
Sagittarius	Sgr	Teapot	ε–γ–δ–ε–ζ–τ–σ–φ–δ–λ–φ–ζ
Scorpio	Sco	Scorpion	β–δ–π δ–σ–α–τ–ε–μ–ζ–η–θ–ι–κ–ν–λ
Taurus	Tau	Bull Horns	β–α–ζ
Ursa Major	UMa	Big Dipper	α–β–γ–δ–ε–ζ–η
Ursa Minor	UMi	Little Dipper	α–δ–ε–ζ–ν–γ–β
Virgo	Vir		β–η–γ–δ–ζ–α–γ δ–ε

Project 1-1
Charts, Stars and Magnitudes

Attached is a finding chart for the brighter stars in the open star cluster named M67 which lies in the constellation of Cancer. Its name indicates that it is the 67th object in the catalog compiled by Messier. Each star is identified by a 2 or 3 digit "ID" number generally written to the right of the star it identifies (but not always).

Visual photometry is the process of determining the brightnesses of an unknown star by comparing its brightnesses to nearby standard stars of known brightness. You make this comparison with your eye. The standard stars that match the unknown closest in brightness set limits on what the visual magnitude of the unknown star can be; if the unknown is fainter than standard A but is brighter than standard B, then its visual magnitude is somewhere between those of A and B.

For this exercise, you will work with the image of a star cluster projected from a slide. Attached is a sheet with standard stars and unknown stars for you to work with. The standard stars are arranged in two ways: in Table 3 they are listed by ID number, in Table 4 by visual magnitude. Use Table 3 when you want to find the visual magnitude of a particular star, use Table 4 when you're trying to find a star of a particular magnitude.

Follow the process described below for the unknowns assigned by your instructor on the attached sheet. When you have finished all the unknowns, record your results in Tables 1 and 2. Be sure to list them from brightest to faintest (small mag to large mag) in Table 2.

Then review your results to see if they are internally consistent. That is, if you say unknown U is brighter than standard A, but unknown V is fainter than standard A, then your results should show unknown U brighter than unknown V. If not, you have problems.

Example

Suppose you want to determine the visual magnitude of star U (a fictional unknown star). First find U on your finding chart, then locate it it on the slide. Next identify a standard that is near U on the slide and on your finding chart, for example A. Compare the two. Which is brighter? Keep track of your observations by making notes like the following:

Unknown U is fainter than standard A of V mag 2.87.

Since you know U is fainter than 2.87 magnitude, now compare it to a standard that is fainter than 2.87 magnitude, like B and record your results:

Unknown U is brighter than standard B of V mag 8.69.

You have now narrowed U's magnitude down to somewhere between 2.87 and 8.69 . Continue to work on narrowing it down. For example, compare it to standard C.

Unknown U is brighter than standard C of V mag 4.3.

You have now narrowed U's magnitude down to somewhere between 2.87 and 4.31. Now compare to standard D.

Unknown U is slightly brighter than standard D of V mag 3.88.

So you now know U is somewhere between 2.87 and 3.88 in magnitude, but probably pretty close to 3.88. Pretty good so far, but the range is still too large. Now, find another standard slightly brighter than D and compare U to it.

Unknown U is slightly fainter than standard E of V mag 3.64.

At this point, you should have narrowed the visual magnitude of U to the 3.64 to 3.88 magnitude range. Now recompare U to both of these stars. Which is it closer to in brightness? We could call it 3.76, if it appears halfway between, or 3.70, if it is closer to the 3.64 magnitude standard, or 3.80, if it is closer to the 3.88 magnitude standard. Such a train of thought looks like;

Unknown U is slightly fainter than standard E of V mag 3.64.
Unknown U is slightly brighter than standard D of V mag 3.88.
Unknown U is closer to standard E.
I guess Unknown U has a V magnitude of 3.70.

Tricks of the Trade

1) Use Table 3 when you want to find the visual magnitude of a particular standard star.

2) Use Table 4 when you're trying to find a standard star of a particular visual magnitude.

3) Don't zero in on your star too fast by visually picking one that at first glance seems to match; you won't be that lucky at the telescope.

4) The last two standards that bracket each unknown should not differ by more than 0.5 magnitudes.

5) Look for small patterns of stars (triangles, squares, pentagons, etc.) to find your way around the slide.

6) The finding chart shows the stars photographed in blue light while the slide is photographed in yellow light. That's why the sizes of a star image on the finding chart don't always match its image on the slide — you are seeing the differences in the colors of stars in black and white.

7) Make your comparisons in the same order each time. Don't compare the unknown to the standard and then the next time compare the standard to the unknown. Yes, there is a subtle but significant difference.

8) Never treat an unknown as a standard — don't use an unknown to determine the magnitude of another unknown.

Questions

1) Approximately how many times brighter is the brightest unknown star than the faintest?

   Hint: Use the table in Box 1-2 to convert magnitude difference to brightness ratio.

2) What property of the star images are you are using to estimate the visual magnitudes of stars on the slide?

   Hint: What you are measuring is their visual magnitude, but what is it that you can see that differs from one star image to another on the slide? What is it you see that you mentally relate to visual magnitude?

3) Why is star 81 so much brighter on the finding chart than the slide, especially compared to stars 108 and 170? What are the colors of stars 81, 108 and 170? Explain.

   Blue stars and red stars of equivalent visual magnitude will look different on photographs taken in different colors of light. A blue star's image on the finding chart (photographed in blue light) will look bigger than its image on the slide (taken in yellow light) while red stars will look brighter on the slide (taken in yellow light), than on the finding chart (photographed in blue light).

4) List and discuss the sources of error (things that give you trouble) in performing visual photometry.

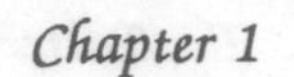

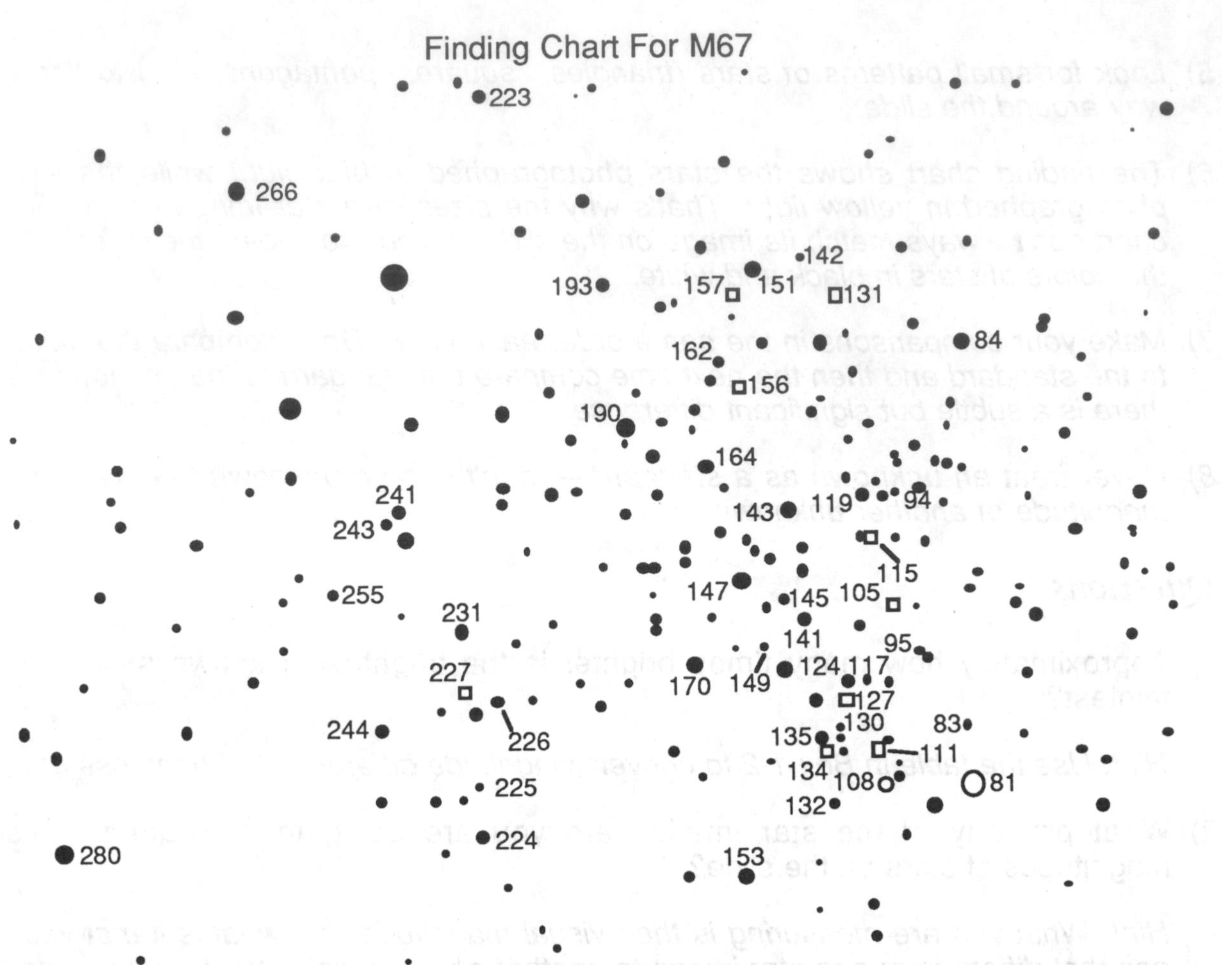

M67 Unknowns

Table 1: Unknowns Ordered by Star

Star #	V Magnitude
81	
105	
108	
111	
115	
127	
131	
134	
156	
157	
227	

Table 2: Unknowns Ordered by V Magnitude

Star #	V Magnitude

M67 Standards

Table 3: Standards Ordered by Star

Star #	V Magnitude
83	13.24
84	10.59
94	12.83
95	12.67
117	12.61
119	12.57
124	12.14
125	13.86
130	12.88
132	13.12
135	11.45
136	11.31
141	10.48
142	14.17
143	11.52
145	12.82
147	13.28
149	12.56
151	10.49
153	11.31
162	12.84
164	10.55
170	9.69
175	13.72
190	10.98
193	12.26
223	10.58
224	10.76
225	13.08
226	12.77
231	11.50
241	12.68
243	12.61
244	10.78
255	12.72
266	10.55
280	10.70

Table 4: Standards Ordered by V Magnitude

Star #	V Magnitude
170	9.69
141	10.48
151	10.49
164	10.55
266	10.55
223	10.58
84	10.58
280	10.70
224	10.76
244	10.78
190	10.98
136	11.31
153	11.31
135	11.45
231	11.50
143	11.52
124	12.14
193	12.26
149	12.56
119	12.57
117	12.61
243	12.61
95	12.67
241	12.68
255	12.72
226	12.77
145	12.82
94	12.83
162	12.84
130	12.88
225	13.08
132	13.12
83	13.24
147	13.28
175	13.72
125	13.86
142	14.17

Project 1-2
Mapping the Heavens

Table 1 lists a number of desired parameters for stars. The first column is for the common name, the second is for the Bayer-Flamsteed name, the third column is for right ascension, the fourth column is for declination, the fifth column for apparent visual magnitude, and the last column for the map number wherein each star is found. Note that for some stars only the Bayer-Flamsteed name is listed, while for others only the right ascension and declination.

Your mission is to supply all the other parameters from either of these two clues. Proceed as follows for each star assigned:

Starting from the Bayer-Flamsteed name:

1) Find the map number that contains the star's constellation in The Trained Eye Star Atlas.
2) Search the constellation for the Greek letter or number of the Bayer-Flamsteed name
3) Note the common name if any.
4) Figure out the apparent visual magnitude from the star's symbol.
5) Interpolate right ascension and declination from the map grid.

Starting from right ascension and declination:

1) Leaf through the atlas for the map that includes the right ascension.
2) Go left to its hour of right ascension line.
3) Go down the hour of right ascension line to the declination degrees and arc-minutes.
4) Go left to the minutes of right ascension and you're there.
5) Note the Bayer-Flamsteed name.
6) Note the common name, if any.
7) Figure out the apparent visual magnitude from the star's symbol.

In both cases, make a small sketch of each constellation showing the star's position in the constellation. For example, Sirius is in the constellation Canis Major. Your sketch will look something like this:

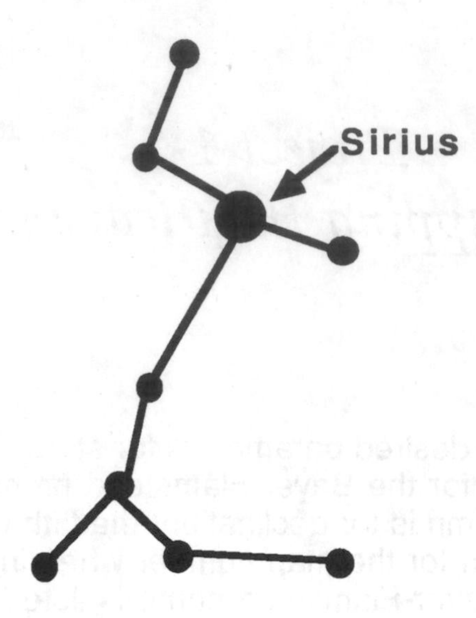

Questions

1) What are the common names, Bayer-Flamsteed names, and apparent visual magnitudes for the brightest star in Orion, Taurus, Auriga, Gemini, Canis Minor, Canis Major?

2) What are the common names, Bayer-Flamsteed names and apparent visual magnitudes for the brightest star in Leo, Ursa Major, Bootes, Virgo?

3) What are the common names, Bayer-Flamsteed names, and apparent visual magnitudes for the brightest star in Lyra, Cygnus, Aquila, Scorpio?

4) What are the common names, Bayer Flamsteed names, and apparent visual magnitudes for the brightest star in Andromeda, Pegasus, Perseus, Cassiopeia?

5) What is the distance between Vega and Deneb, Deneb and Altair and Altair and Vega in degrees?

   Hint: Measure the distance between two stars in millimeters. Then measure along the declination scale that same number of millimeters, starting from zero degrees declination to convert the millimeters to degrees.

6) What is the magnitude difference between the brightest and faintest stars from Table 1 and your responses to questions 1,2,3 and 4? Approximately how many times brighter is the brighter star?

   Hint: See Box 1-2.

7) Do stars become more numerous as you go to fainter magnitudes or to brighter magnitudes?

   Hint: Tablulate the number stars with magnitudes less than 0, between 0 and 0.99, between 1.00 and 1.99 to support your answer.

Table 1: Bright Stars

Common Name	Bayer-Flamsteed Name	RA H	RA M	DEC °	DEC '	Apparent Visual Magnitude	MAP #
	α And						
		1	38	-57	14		
	α UMi						
		2	08	23	32		
	α Tau						
		5	14	-8	12		
	α Aur						
		5	26	6	21		
	β Tau						
		5	55	7	24		
	α Car						
		6	45	-16	42		
	ε Cma						
		7	40	5	14		
	β Gem						
		10	09	11	59		
	α Cru						
		12	47	-59	41		
	ζ UMa						
		13	26	-11	10		
	β Cen						
		14	15	19	12		
	α Cen						
		16	29	-26	26		
	λ Sco						
		18	37	38	47		
	α Aql						
		20	42	45	17		
	α Psa						

Chapter 2
The Mortal's Point of View

The star maps in The Trained Eye Star Atlas show the celestial sphere from the inside, but in celestial coordinates, not the coordinate system we mortals live in. After all, when was the last time your head pointed north? Unless you're an astronaut or an Eskimo, the answer is never. We live in *human coordinate systems* where our heads point *up* not north. Still we constantly confound both coordinate systems by using phrases like "up north" and "down south" and our eternal habituation to hold maps north side up even when looking south. If you go outside and hold The Trained Eye Star Atlas north side up, it won't guide you through the stars. Only when you hold The Trained Eye Star Atlas in celestial coordinates does it work. Not as onerous a task as you now think if you just approach it with the right slant.

Human Coordinates

Your *human coordinate system* starts with *up*, something you've learned intuitively by age one in standing up and falling down — otherwise known as discovering gravity. The Earth's gravity defines which direction your head points in the universe. That point straight overhead is your *zenith*, and the point straight underfoot is your *nadir*. From these two fundamental points, you can define the rest of your human coordinate system. That's right, your very own coordinate system, for nobody else on the Earth has the same zenith and nadir as you. From zenith and nadir, you can now define up and down.

Point a finger at your nadir (feet). Swing your arm from your nadir to your zenith (overhead) and it moves *up.* Swing your arm from your zenith to your nadir and it moves *down.* Now point your finger halfway between your zenith and nadir and spin around in a circle. Your finger sweeps out a circle about you called your *horizon.* Notice that your horizon lies halfway between your zenith and nadir, not where the sky meets the ground (fall into a well and the sky meets the ground near your zenith, not at your horizon). Up and down give directions between zenith and nadir but what about directions around the horizon?

The two remaining directions in your human coordinate system, *left* and *right*, are not so intuitive. To define them, we have to rely on something else, so take your watch off and place it on the ground at your nadir. Point a finger outwards at your horizon and watch your watch. As you turn *clockwise*, your finger will swing around your horizon to the *right*. Turn *counterclockwise* and your finger swings around your horizon to the *left*.

The horizon also gives the official name to your human coordinate system; the *horizon coordinate system.* It shall remain human coordinate system in this book. You can finish your human coordinate system by defining coordinates and giving them names and measures. The *altitude* coordinate measures the distance, in degrees (°), arc-minutes ('), arc-seconds(") up or down from your horizon. Altitude starts from zero at your horizon and goes up to +90 ° at your zenith or down to -90 ° at your nadir. It's coordinate pair *azimuth* measures the distance rightward around your horizon in degrees (°), arc-minutes ('), arc-seconds("). Azimuth starts at *zero* degrees from a magic point on your horizon called the *north point* and increases rightward around the horizon *360 °* back to where you started from. What, however, is the north point?

Look north from where you stand (that's Earth north like on a compass or a road map). The Earth's surface curves down away from you, so you can't see all the way to the north pole. Where your gaze dead-ends at the horizon is your *north point,* the furthest point north you can see (Figure 2-1). Look to the right of your north point for your *east point* at an azimuth of 90 °. Another 90 ° further brings you to the *south point* at an azimuth of 180 °. Swing just 90 more degrees on around and there's your *west point* at azimuth 270. Look at a compass and you'll see all these points engraved right on it.

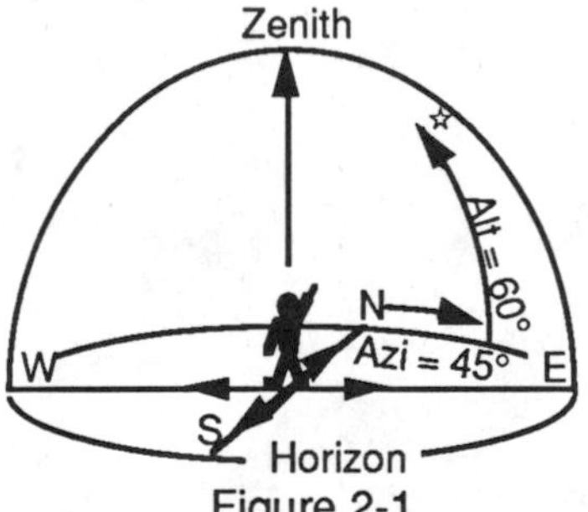

Figure 2-1

Using the coordinate pair altitude and azimuth you can pinpoint the location of any object in the little piece of the universe that extends from the earth at your feet above the horizon to infinity overhead. Your human coordinate system provides the framework within which your mind structures the universe. Without a structure, you cannot interpret what your eyes see. Objects are indistinguishable, for they are all just "out there." Without the structure of positions and directions provided by your human coordinate system, you would be paralyzed. Which direction do you move? Where do you reach?

Box 2-1: Perception and Night Vision

You walk into a theater and everything goes dark. You stop and, in impatient expectation, wait for sight to return — blind to why it failed in the first place and the miracle that it does recover. Yet your sight does return in a process called *dark adaptation.*

Light gets into your eye through your *pupil,* the adjustable hole provided by your *iris.* Your *lens* focuses this light to produce an image on your *retina.* Your retina contains millions of light-sensitive cells, called *rods* and *cones,* which convert light stimulus to electrical sensation that flows up your *optic nerve* to your brain. You owe your *color vision* to the cone cells, your *low light level vision* to your rod cells.

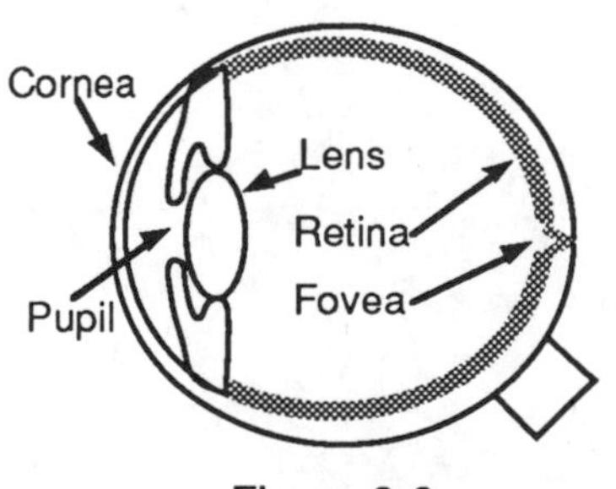

Figure 2-2

In the center of your retina lies a small dimple called the *fovea centralis,* where light-sensitive cells are packed together more tightly than on the rest of your retina. You read with your fovea centralis, for it's the place on your retina, thanks to all the tightly packed detectors, where your vision is sharpest. As you read this sentence focus on one word and see how many words to either side of it you can make out without moving your eye. There's nothing wrong with your eye if you can't read more than one word either side; it merely reflects the economy of nature. You don't need *high acuity* of vision everywhere on your retina, just one spot that you move around to where you need it.

Not only are there more detectors packed into the fovea centralis, most of these are cones. Elsewhere on the retina rods predominate. Thus, your color vision is most vibrant at the center and progressively washed out to the edges.

Your iris provides the first line of defense against fluctuating light levels. When light is plentiful, it contracts your pupil to keep your retinas from burning out. As the world darkens, your iris dilates your pupil in compensation. Of course, your pupil cannot dilate forever. When it reaches its maximum size, you finally perceive the world growing darker and your retina starts a slow chemical and neural change to increase its sensitivity. The chemical change involves the increase of the photochemicals in the rods and cones as light decreases. These photochemicals make your rods and cones work; the more they have the more sensitive they are.

Your eye also makes some neural trade-offs. As you become dark adapted, you don't see detail quite as well; your retina is adding stimulus from several adjacent rods to produce a single sensation. Your eye also increases the time over which they add up photons before generating a sensation; much like photographers using longer exposure times to compensate for decreased light levels.

These changes take about 7 minutes for the cones and up to an hour for the rods. Although both become more sensitive, the rods start and end far more sensitive than the cones. So at night you see with your rods and not your cones (which explains why you don't see colors at night).

The end result of the whole dark-adaptation process is that your eyes, over an hour, become approximately 10000 times more sensitive to light. Now the big question: how long does it take to destroy dark adaptation? Just one flash of bright white light and you have to wait another hour. So what do you do if you want to read in the dark?.

Use a trick of color. Your rods, your low-light-level sensors, don't see red light. Use a red flashlight, and your rods don't know its on and don't switch back to bright-adaptation. Your cones, however, do see in red light, so you use them to read. This also explains all those old submarine movies where they always use red light in the conning tower when surfacing at night.

Now where do you get a red light? You can convert a flashlight to red using red cellophane, red nail polish, brown paper bags, etc. The best way, however, is to start with red light. Certain solid-state electronic devices, called light emitting diodes (LED), emit pure red light, never burn out and are up to 8 times more energy efficient than incandescent bulbs so batteries last a lot longer.

By Any Other Name

Reach for the stars if you know where. Although the human and celestial coordinate systems seem very different they are structurally identical (Table 2-1). Both coordinate systems are *spherical coordinate systems* (maps of spheres) and both map the same sphere (the sky). Each system has two fundamental points, one great circle halfway between the fundamental points, four directions and two coordinates. The fundamental difference is how the Earth defines the fundamental points; the Earth's rotation defines north and south celestial poles, the Earth's gravity defines zenith and nadir.

When a sphere like the Earth spins, everything moves except two special points (Figure 2-3). Call these two points *poles* and name one the *north pole* and the other the *south pole* to avoid confusion. These poles originate the *terrestrial* (Earthly) *coordinates* of *north pole, south pole, equator, latitude, longitude, north, south, east* and *west* as we normally use them on the Earth, not the celestial sphere. The north celestial pole is that point on the celestial sphere directly over the north pole of the Earth, and the south celestial pole is directly under the south pole.

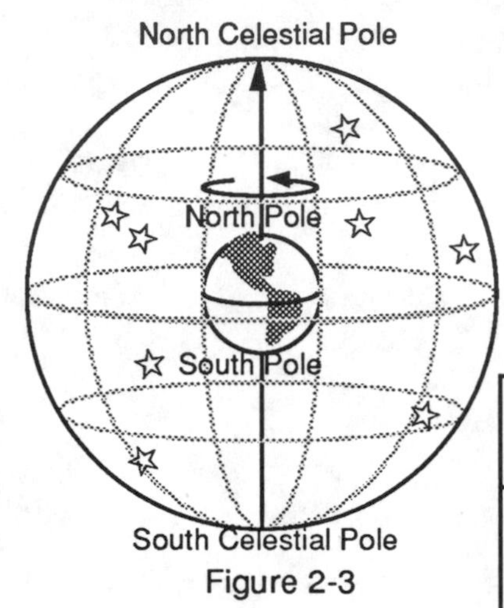

Figure 2-3

Table 2-1: Coordinate Systems Structures

Structure	Celestial	Human
fundamental points	north celestial pole	zenith
	south celestial pole	nadir
defined by earth's	rotation	gravity
great circle	celestial equator	horizon
directions	North	up
	South	down
	East	left
	West	right
coordinates	declination	altitude
	right ascension	azimuth

Gravity defines your zenith and hence all your human coordinate system except for the north point. That requires the Earthly direction north (defined by the Earth's rotation). So your human coordinate system depends upon the Earth's rotation, as does the celestial coordinate system — both systems are tied to the Earth. The Earth provides the bridge from one to the other. Referring back to the issue that started this chapter; how do you hold up The Trained Eye Star Atlas? You should really ask: "How do I transform the celestial coordinate system into my human coordinate system?"

Between Heaven and Earth

The best place to start is where both coordinate systems look the same, so put on your parka and start walking to the north pole. At the north pole, the Earth's gravity aligns you perfectly with the Earth's rotation axis (since you're standing right on top of it). The north celestial pole is directly over the Earth's north pole at your zenith, the celestial equator lies at the horizon. Here and here alone directions in celestial coordinates match common sense: up is north, down is south, right is west, left is east (Figure 2-4). Common sense also cautions you about frostbite, so start hiking south but walking backwards so you can keep an eye on the zenith and north celestial pole.

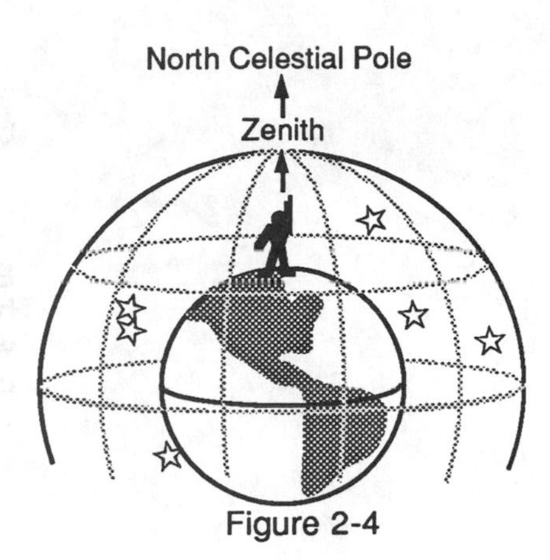

Figure 2-4

One step south and the north celestial pole no longer hovers at your zenith since you no longer align with the Earth's rotation (Figure 2-5). The further you hike back over the curve of the Earth, the further your head tilts away from the north celestial pole. However, being human and thinking the universe revolves around you, you perceive the celestial pole as tilting away from your zenith.

Look over your shoulder in the direction you're walking and you'll see the celestial equator tilting upward just as fast as the north celestial pole drops. Look to the west and you'll see the celestial equator pivoting clockwise around the point where it touches the horizon. When you're two thirds of the way back to the Earth's equator (at latitude 30°), stop to take off your parka and look around to orient yourself (Figure 2-6). Back north, where your footprints fade away into the distance, the north celestial pole has dropped 60 ° down from the zenith to just 30 ° above the horizon (Figure 2-6a). An interesting coincidence by the way, you can determine your latitude by measuring the altitude of the north celestial pole. Off to the west and east, the celestial equator tilts up at a 60 ° angle from the horizon. Although the celestial sphere has tilted, it has not distorted (Figure 2-6b).

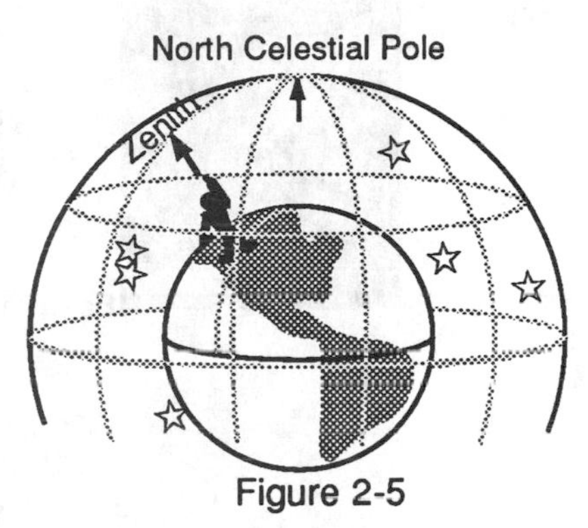

Figure 2-5

Now you know how to transform the star maps in The Trained Eye Star Atlas so they match the sky; just tilt them. Then the positions of stars on the map match what you, in human coordinates, see in the sky. The Trained Eye Star Atlas guides you through the stars. Of course, The Trained Eye Star Atlas maps the celestial sphere in sections, not as one big ball, so determining how to tilt and place each star map overhead takes a bit more thought.

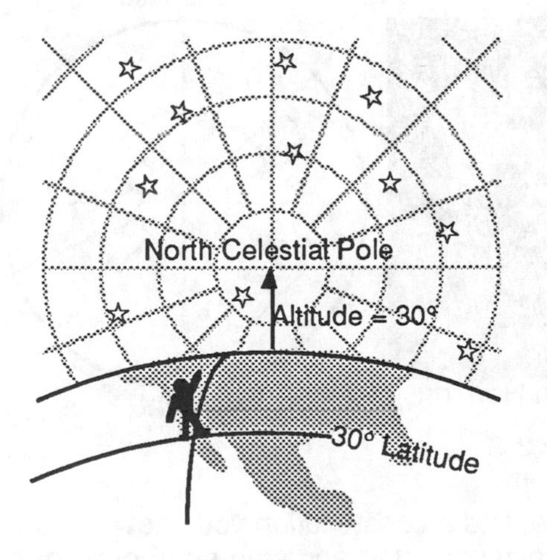

Figure 2-6a

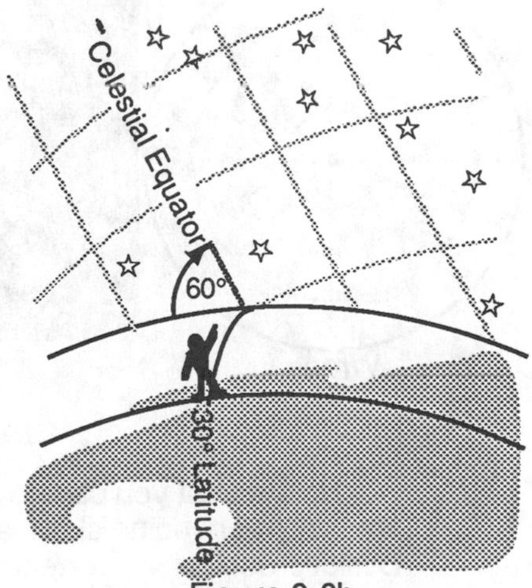

Figure 2-6b

Imagine a celestial compass needle running up the middle of each equatorial map in The Trained Eye Star Atlas with N at the top and S at the bottom. Hold up each star map with the compass needle always pointing to the north celestial pole in the sky. In different parts of the sky, the map may be upside down or laying on its east or west side, but in all cases it will match what you see in the sky (Figure 2-7). Keep in mind that only at the Earth's equator (where the north celestial pole will have tilted down to the horizon and the celestial equator passes straight overhead from east to west) will an equatorial map lie flat on its side to match stars at the eastern or western horizon.

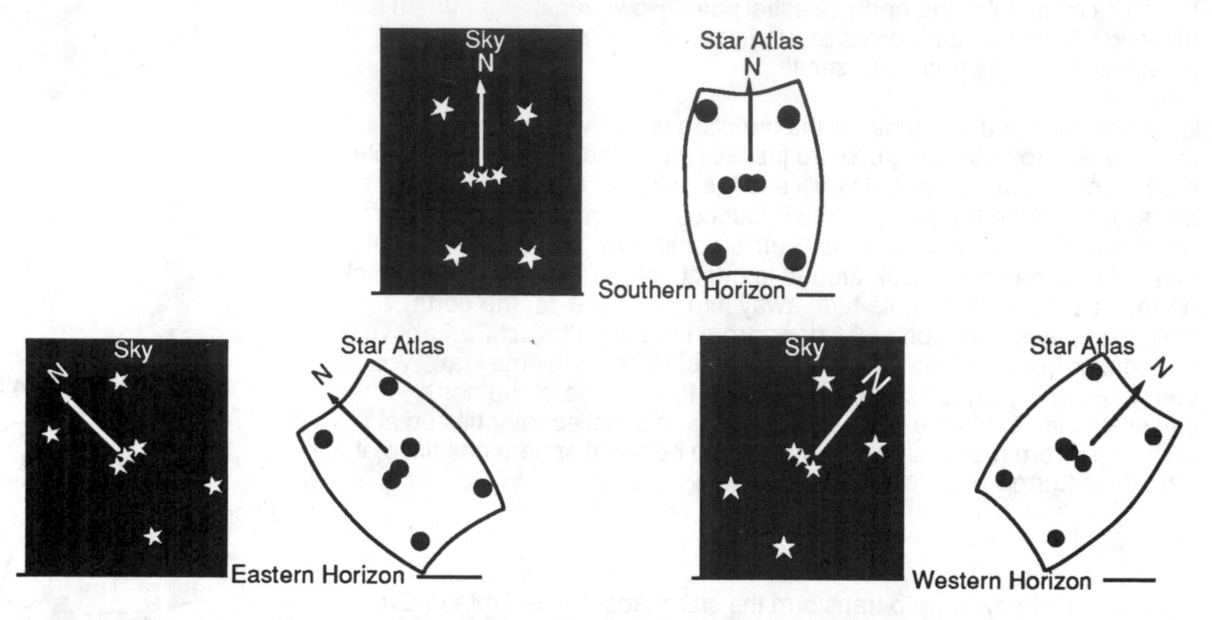

Figure 2-7

Polar maps are easier to orient properly. Hold the appropriate one with the celestial pole on the map centered on the celestial pole in the sky. Then just rotate the map around its celestial pole until a prominent constellation lines up which lines up all the other constellations (Figure 2-8).

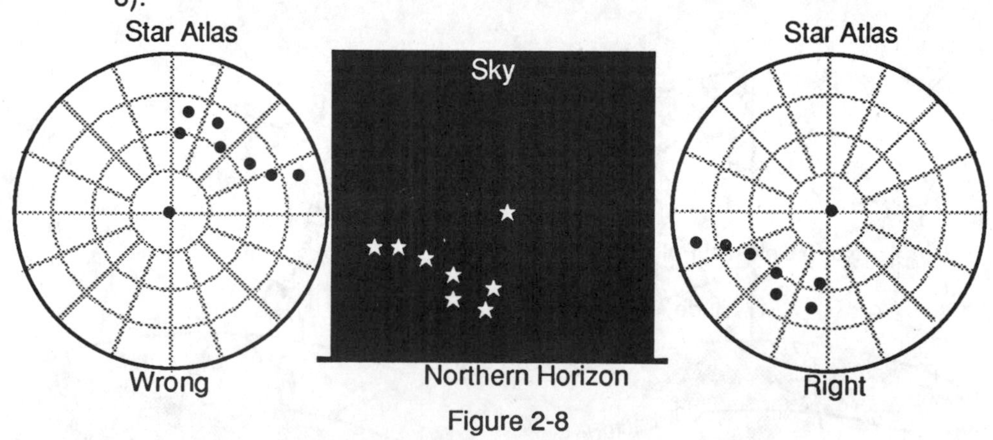

Figure 2-8

Now at last you can go outside, find a constellation you know, hold up its corresponding star map correctly oriented and start star hopping. Besides just looking correct, the star map tells you which way and how far to go to get from one star to another. The patterns you see on the star map show you the patterns in the sky. If two stars on a star map line up to point at a third, then look for them to do the same thing in the sky. If the distance between two stars on the map is 10°, then the distance between them in the sky will be about 10°. You can use the declination scale to estimate distances between stars on each star map, but how do you estimate degrees in the sky? Use the following handy table .

Now at last you can go outside, find a constellation you know, hold up its corresponding star map correctly oriented and start star hopping. Besides just looking correct, the star map tells you which way and how far to go to get from one star to another. The patterns you see on the star map show you the patterns in the sky. If two stars on a star map line up to point at a third, then look for them to do the same thing in the sky. If the distance between two stars on the map is 10°, then the distance between them in the sky will be about 10°. You can use the declination scale to estimate distances between stars on each star map, but how do you estimate degrees in the sky? Use the following handy table .

Table 2-2 Handy Angles

Hand	10°
Finger	1°
Pinky Finger Nail	0.5°

Hold your hand up at arm's length. Your pinky finger nail covers about 1/2° of the sky, your pinky finger is about 1° wide and the width of your hand with fingers and thumb together spans about 10°. The moon is about 1/2°, and you can cover it with your pinky finger nail. Try it and you'll see just how tiny the moon is compared to how big it seems.

At last The Trained Eye Star Atlas makes sense as a guide to the stars in human coordinates. Still, reflect on one thing: human coordinates cover only half the celestial sphere - the half above your horizon that you can see. You know there are more stars than that — after all ,The Trained Eye Star Atlas maps a full celestial sphere's worth. So how and when do you see the other half of the stars?

Box 2-2: Where Have All the Stars Gone?

Once you identify the seasonal sky, you can use your star maps to find other constellations. If you have trouble finding a particular star, it helps to distinguish lack of knowledge from circumstance. How do you determine whether you can't find a star because it's not up or because you don't know where to look? Here's how to use The Trained Eye Star Atlas to prove whether a star is not up and even know where it is not up.

By trivial definition, a star that is *not up* is *not above the horizon.* Find a star map in The Trained Eye Star Atlas that matches the constellations just above the horizon and hold it correctly oriented (north end pointed towards the north celestial pole). Identify the lowest star you can see in the sky and find its position on that star map. Then, if you are holding the star map correctly oriented, the horizon corresponds to a horizontal line across the map just below that star on your map.

If only life were that simple. Most likely if you live in a light-polluted environment, the lowest star you can see will be 20° (measured by using hands at arms length) above the horizon. That's the clue. If the horizon is 20 degrees below that star in the sky, it will also be 20° below that same star on the map. So identify that star on your star map, mentally draw a horizontal line 20° below it on the star map and that is your horizon on the star map. Using the declination scale on the map you know how many millimeters correspond to 20°, so that will tell you how far below the star to draw the line. Once again, remember that the horizon will be canted across the page since the celestial sphere is tilted over with respect to your real horizon.

Once you have the horizon drawn on the map, any star below it is down, any star above is up. The altitude of any star below the horizon equals its distance in millimeters below the horizon on your star map converted back to degrees with a minus sign (-) tacked on to indicate below the horizon. Its azimuth equals that of any star directly above it. So pick a star you can see more or less directly above it on the map and use that as the azimuth.

You can do this same procedure for any star just barely above the horizon to determine where to look for it, and if you still can't see it, then at least you know to blame it on something besides the horizon.

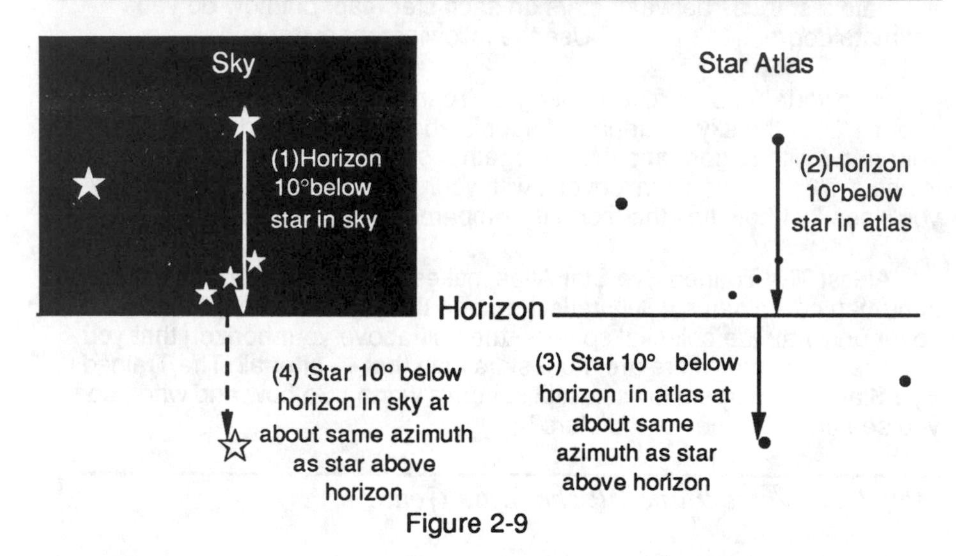

Figure 2-9

The View from a Spinning Earth

As you read this page, how much of your surroundings do you see? What's that behind your head? As you rotate your head around to look, think of the Earth's rotation spinning you around inside the celestial sphere. You never see more than half of the room without rotating your head; you cannot see more than half of the celestial sphere without the Earth rotating. Your head provides an excellent model for understanding the Earth's *diurnal* (daily) *rotation*; the rotation of the Earth and how it causes stars to rise in the east, cross the sky and set in the west.

Using your head as a model Earth, hold your head still and see how far up, down, left, and right you can see. Your field of view (how far you can see around your head without rotating it) ends at your eyebrows, cheek, nose, etc. This corresponds to where the sky meets the Earth in your human coordinate system — the horizon (not the correct definition of horizon, but close enough unless you're sitting at the bottom of a well). Now pretend that the objects around you are stars on the celestial sphere and watch how they rise (pop into view), cross your vision and set (disappear from view) as you rotate your head. It is impossible, as you know from Chapter one, to draw the celestial sphere on a flat sheet of paper — this page — all at once without distortion. So Figures 2-10, 11, and 12 present their views from the inside in four vignettes; what you see looking to the northern horizon, what you see looking to the easter horizon, what you see looking to the southern horizon, and what you see looking to the western horizon.

Look straight overhead to simulate the view from the Earth's north pole. With your head reclined at an uncomfortable angle, your eyes look out along your axis of rotation — your zenith is at the point straight overhead, your north celestial pole, which sits still as you rotate counterclockwise. All the stars in your field of view circle around your north celestial pole, neither rising nor setting (Figure 2-10). Since the stars never set then stars below the celestial equator (which, as you remember, lies on your horizon when your standing at the north pole) never rise. You see just half of the celestial sphere, forever swiveling about overhead.

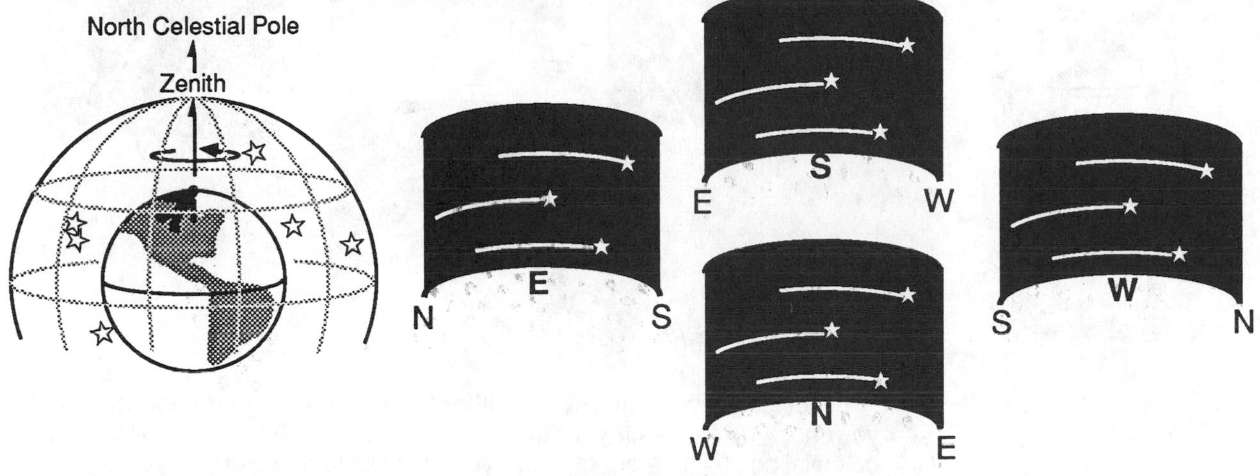

Figure 2-10

Give your neck a break and sit upright to simulate the view from the Earth's equator. Now your north celestial pole lies somewhere up in your eyebrows and your south celestial pole down in your cheek, and your zenith points straight out towards the celestial equator. As you swivel counterclockwise in your chair, stars zip clockwise straight across your field of view (Figure 2-11) and you eventually see all the sky as you spin

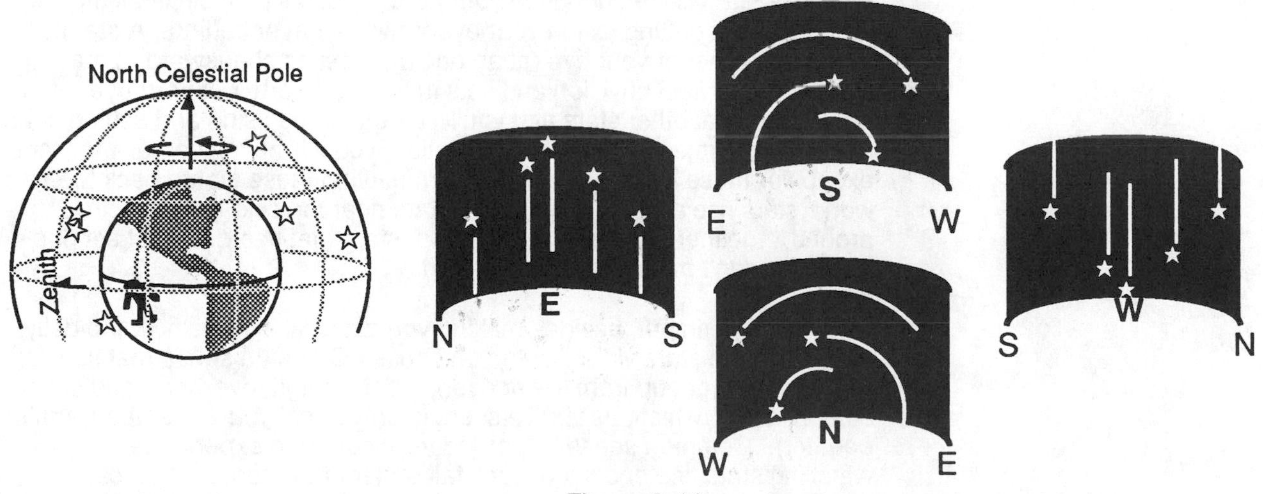

Figure 2-11

around. Slow down and watch things more closely. A star that first appears in the left corner of your eye will rise straight up, cross the center of your vision and set straight down in the right corner of your eye. Look up to your eyebrows and try again. A star pops straight up out of a left eyebrow ,crosses above to the top of your field of view and sets straight down in a right eyebrow. Translating this back into the view from the Earth's equator, stars that rise due east cross directly overhead and set due west; all stars rise straight up at the eastern horizon and set straight down at the western horizon.

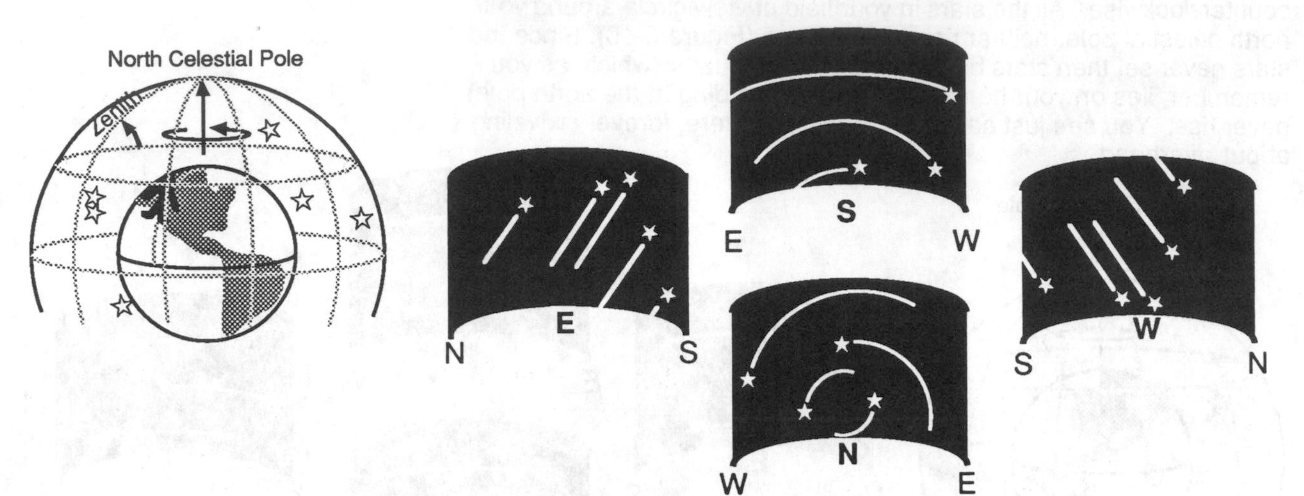

Figure 2-12

Most likely you don't live at either the north pole or the equator ,so tilt your nose up to match your latitude on the earth. As you do, your north celestial pole climbs out of your eyebrows (as fast as you tilt your nose) while your south celestial pole sinks below your chin (Figure 2-12), as do the stars near it. Those stars near your south celestial pole never rise, and those stars near your north celestial pole never set. In the real world, this means that from your latitude you will always see some of the celestial sphere and some you will never see. Now start your diurnal rotation by pivoting fully around .

Your north celestial pole sits stationary while all the stars pivot around it. Those stars close enough to your north celestial pole circle all the way around, never getting lost in your eyebrows — never setting. A star rises in the left corner of your eye (at an angle), crosses cheekward of the center of your field of vision and sets in the right corner of your eye (at an angle). Look at other stars and you'll see the same rising and setting at an angle. Each time you complete a rotation, you will have seen all the stars except for those below your cheek. Translating these sights back the real world, stars rise and set at an angle, stars near the celestial pole circle around it ,neither rising nor setting, and stars that rise due east set due west, but don't pass straight overhead.

This experiment, besides making you dizzy, illustrates how the daily motions of the stars violate common sense. Common sense that says stars rise straight up from the horizon, cross straight overhead and set straight down (which, as you've seen, is only true if you live at the Earth's equator). Common sense is learned from common experience, and watching stars rise and set doesn't fall under this category. You can understand the motion because you can model it very simply as shown in this series of swivel-chair experiments — retrain your trained eye. With this model of common experience ,you can now go out and understand the risings and settings of stars.

Diurnal rotation, as you've seen, will eventually bring around all of the celestial sphere (visible from your latitude). Obviously you won't be able to see all the stars because nighttime doesn't last forever. The sun eventually gets in the way. This brings up another question. If the sun blots out the celestial sphere when its up, how do we know what stars lie on the celestial sphere near the sun?

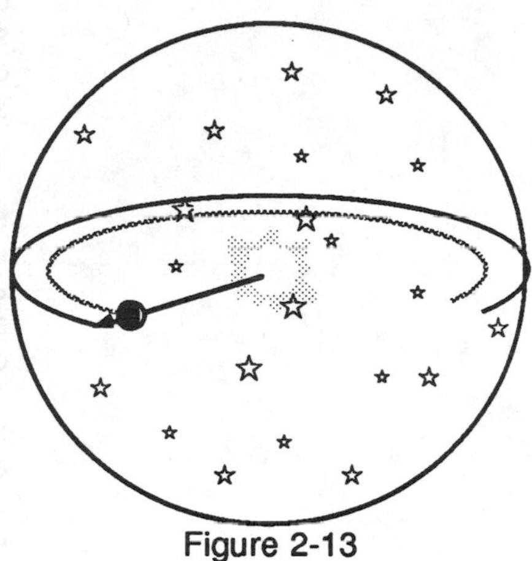

Figure 2-13

To Everything There Is a Season

The sun does not sit in just one place blotting out all knowledge of a particular patch of sky; otherwise the star maps in your star atlas would have a large blank area. The Earth, in addition to rotating on its axis once a day, revolves in an orbit around the sun once a year as you can simulate very easily. Pick a lamp for your pretend sun, and put it in the middle of the room. Stand a few feet away, facing outwards, and walk in a circle around your sun while noticing what you see on the walls (your stars). Stop when you've completed one revolution about your sun. How did you know when to stop? When you again see the same constellations (Figure 2-13), you have completed one orbit around your sun.

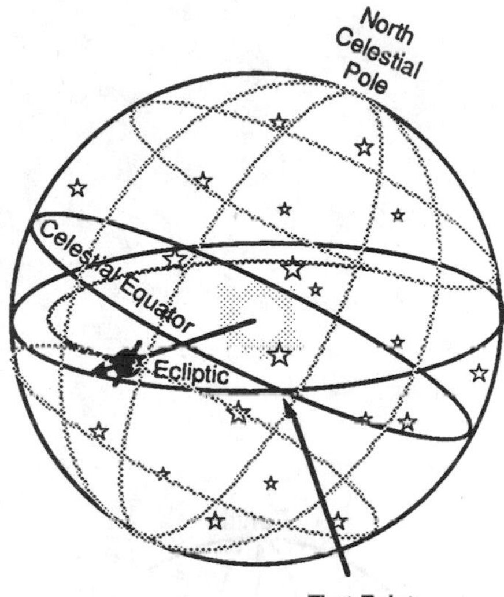

Figure 2-14

You can simulate the combined effects of diurnal rotation and annual revolution by spinning counterclockwise once for each step around your sun. On each spin you'll see your sun rise, cross your sky and set (diurnal motion). When you're looking directly away from the sun during each step, you'll see a slightly different view of the stars. Step and spin enough times and you return to your starting position and sky.

Perform the same experiment outside and you see the Earth revolve around the sun. Go out each evening at the same time (say 10:00 pm) and note which star on your star chart is near your zenith Then note it on your star chart. Come back one diurnal rotation (day) later and a different point on the celestial sphere will be on your zenith (slightly to the east of the previous one). Mark it and repeat the exercise each night. Of course, the night-to-night marks will fluctuate back and forth since the daily change is small (about 1°, the width of your pinky finger at arm's length) and hard to measure accurately. Still, over weeks, you will see your marks marching through the star maps back around to the original mark one orbital revolution (year) later.

Another way to see the revolution of the Earth around the sun is to follow the *apparent* path of the sun through the stars in reflection of the Earth's *real* motion about the sun. This path, the *ecliptic,* is distinct from the celestial equator because unlike the experiment you performed above, the rotating Earth revolves around the sun tilted over 23.5°. So lean over 23.5° as you spin and step around your pretend sun (Figure 2-14). Once again on each spinning step you'll see your sun rise, cross your sky and set, but the motion won't be identical on each successive spinning step. Starting when your head is tilted away from your sun, you'll see it cross low across the sky. However ,on each successive spinning step, it will cross progressively higher in the sky until, when your head tilts towards your sun, it will cross high in your sky. Keep spinning and stepping until your head again points away from your sun and it again crosses low across your sky. You will see that the sun oscillates north and south of the celestial equator as it drifts eastward among the stars. You can perform this experiment outside by watching where the real sun sets/rises.

If the sun did not drift north and south of the celestial equator, then the sun would always rise due east and set due west — life would be eternal spring. Since you can see it's setting point shift back and forth along the horizon over the course of a year, then it must be changing it's declination — the ecliptic must be inclined to the celestial equator. Not only is this annual north/south drift of the sun on the celestial sphere easily observed (Project A-1), it is even more easily felt in the changing of the seasons. For the northern hemisphere, when the sun is north of the celestial equator, it rises early, stays up later and crosses higher in the sky. The season is spring or summer. When the sun is south of the celestial equator ,it rises later, sets early and crosses the sky low to the south. The season is fall or winter.

Besides just giving us seasons, the ecliptic in conjunction with the celestial equator establishes the origin of right ascension. The intersection of the ecliptic and celestial equator where the sun crosses south to north (Figure 2-14) is the "*first point of Aries*" (Chapter 1), so named because this intersection of the ecliptic and celestial equator lies in the constellation of Aries — or did some 4000 years ago. Because of precession (the Earth wobbling as it spins), the celestial equator slips westward along the ecliptic one constellation every 2000 years or so. Presently ,the "first point of Aries" is in the constellation of Pisces heading towards the constellation of Aquarius. Contrary to popular culture, "the Age of Aquarius" will not "dawn" until the 26th century (according to the official constellation boundaries established by the International Astronomical Union in 1930).

A Time to Every Purpose Under Heaven

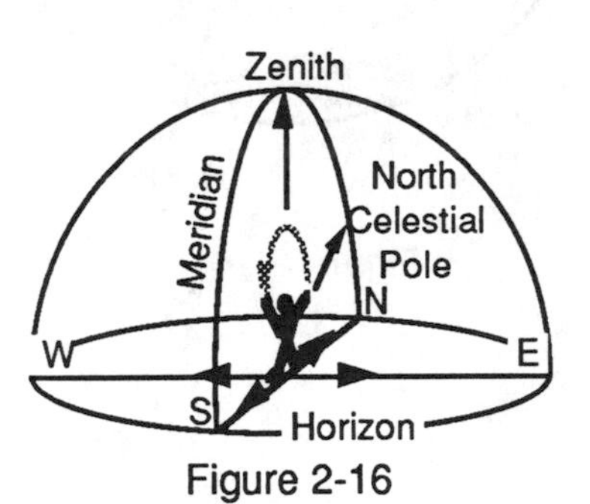

Figure 2-16

Day and year mark the regular and repeated cycles of the earth rotation and revolution. It is for human convenience that we adopt these regular, dependable cycles of nature to mark the passage of time — to supply our clocks. We forget the literal definition of day is "one rotation of the Earth" and year means "one revolution of the Earth about the sun." Depending upon which celestial cycle you wish to monitor against which reference frame, you can have different kinds of days or years — different kinds of time. The classic reference for determining the day is a line across the sky called your *local meridian.* Point at the north celestial pole and swing your finger through your zenith to the south point on the horizon. The line in the sky your fingertip sweeps across the celestial sphere (which divides the sky into eastern and western halves) is your *local meridian* (Figure 2-16).

The first kind of day-time, called *solar time*, clocks the race of the sun from when it crosses your *local meridian* until it comes back around one *solar day* later. Alternatively ,you can define a *sidereal* (stellar) *time* by tracking the first point of Aries as it cycles around the sky. Go out some evening, notice when the first point of Aries is on your meridian (a neat trick) and wait till it comes back around one *sidereal* (stellar) *day* later. The distinction between *solar* and *sidereal time* arises because the two rotations aren't equal. Since the Earth revolves about the sun, the sun appears to drift eastward along the ecliptic about 1° a day. As a result, you have to wait for the earth to rotate about 1° beyond a *sidereal day* (

about 4 solar minutes) for your local meridian to catch up with the eastward drifting sun (Figure 2-17). So the *solar day* is about 4 solar minutes longer than a *sidereal day* (note the distinction of *solar minutes* from *sidereal minutes*). Still, why bother about two kinds of days and times?

You bother with these two types of time because they have different uses. We all live to *solar time*. We sleep, eat, work and play to the cycles of the sun, not the cycles of the stars. *Sidereal time* is the time of astronomy, for it connects celestial coordinates to human coordinates. If you know the sidereal time, you know what right ascension is on your local meridian. If you know what right ascension is on your local meridian, then you know the orientation of the celestial sphere (Figure 2-18) in human coordinates. You know where to look in the human coordinate system for the stars.

Just as we monitor different celestial objects to define the day ,we can monitor different celestial objects to measure the year. The *tropical year* measures the Earth's revolution about the sun *relative to first point of Aries*. Another kind of year, called the *sidereal year*, measures the Earth's revolution about the sun *relative to the stars*. But aren't the positions of the stars defined by the first point of Aries? True, but due to precession (Chapter 1), the first point of Aries slips slowly westward along the celestial equator. As the zero point changes, so do the celestial coordinates for stars. The year measured with respect to the stars, called the *sidereal year*, ignores precession and as a result is about 20 minutes longer than the *tropical year*. The real importance of the distinction between the two types of years is that the tropical year is the year of the seasons — the year of agriculture — locked to the first point of Aries. The sidereal year is the year we can see (the stars visible, the first point of Aries invisible) — convenient to observe but having only a passing association with the seasons.

The ancient Egyptians used the sidereal year by monitoring the *heliacal rising of Sirius*, the first day that Sirius could be seen rising just before the sun, to establish their calendar and predict the flooding of the Nile river. All life in Egypt followed the ebb and flow of the annual flooding of the Nile. Innundation, emergence and drought established their seasons of agriculture. However, the calendar based on the sidereal year drifts from the year of the seasons by 1 day every 70 years. Not much in the lifetime of a human but significant in the lifetime of a civilization. Over the intervening 4000 years, the heliacal rising of Sirius has drifted several months away from the flooding of the Nile. The seasons do not keep pace with the stars.

The Hierarchy Of The Heavens

Now you've seen everything that affects the celestial sphere's appearance: latitude, time and season. How the stars move across the sky depends solely upon your latitude: the north celestial pole sits at its prescribed altitude above the northern horizon, the circumpolar stars circle about it, and other stars following their appointed arcs across the sky. Given enough time you'll eventually see all the stars visible from your latitude rise and set. But whether

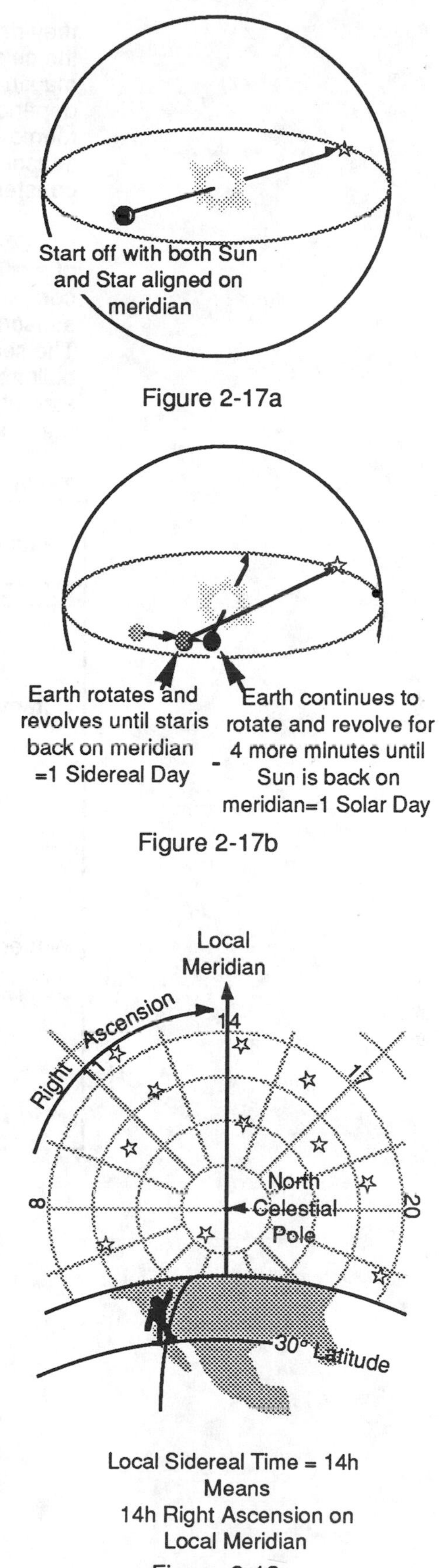

Figure 2-17a

Figure 2-17b

Local Sidereal Time = 14h
Means
14h Right Ascension on
Local Meridian

Figure 2-18

they rise during the day or at night depends upon where the sun is on the celestial sphere — upon the season of year. Knowing which star map in The Trained Eye Star Atlas matches the stars you see likewise depends upon your latitude, time and season. Before you go out and memorize the location of each constellation for every time of day and season of year, try adding one final level of structure over the constellations.

Divide the celestial sphere into four seasonal skies each composed of a different set of bright constellations (Table 2-3). Each seasonal sky corresponds to the constellations overhead in early evening during that season (as the night progresses, you'll see the two seasons that follow). The seasonal skies are easy to recognize thanks to the seasonal patterns built from the brightest stars. These seasonal skies give you a place to start at the beginning of night. From them and your star atlas ,you can find the other constellations, and from them, individual stars.

Table 2-3: Seasonal Skies

Season	Constellations	Bright Stars	Seasonal Pattern
Spring	Leo	Regulus	Follow the arc of the
	Ursa Major		.handle of Ursa Major to
	Bootes	Arcturus	Arcturus, Speed onto
	Virgo	Spica	Spica
Summer	Scorpio	Antares	
	Sagitarius		
	Lyra	Vega	Summer Triangle
	Cygnus	Deneb	composed of Vega,
	Aquila	Altair	Deneb and Altair
Fall	Pegasus		The Great Square of
	Andromeda		Pegasus
	Cassiopeia		
	Perseus		
Winter	Orion	Betelgeuse	Winter Triangle
		Rigel	composed of
	Taurus	Aldeberan	Betelgeuse, Procyon
	Auriga	Capella	and Sirius
	Gemini	Castor	
		Pollox	
	Canis Minor	Procyon	
	Canis Major	Sirius	

Project 2-1
The Night Sky

At last you're ready to go outside and look at *real* stars! While you're at it, measure the *altitude* (the height of the star from the horizon measured in degrees, remembering that the zenith is at 90°, the horizon at 0°), *azimuth* (the angle along the horizon measured clockwise from the north point, in degrees from 0 to 360), *color* (blue, white, yellow, orange, red) and *time* (the time of night you made your observation) .

First, find the point on the horizon directly below the star by pointing your finger at the star and dropping your arm down to the horizon. Altitude is just the number of hands (held at arm's length) up from that point on the horizon to the star multiplied by 10. Azimuth is just the number of hands (held at arm's length) rightward around from north to that point on the horizon directly below the star multiplied by 10. Of course, measuring hands all the way around from north can lead to big errors. So why not measure the azimuth distance from due east, due south, or due west (whichever is closest) and just add the azimuth of due east (90), due south (180), or due west (270) to get the answer.

1) Observe the stars assigned by your instructor in Table 1.

2) For the stars you can't find, explain where they are. Using the technique described in Box 2-2, you can determine if a star is below the horizon and estimate where and when it will rise. If you deduce that it is up then determine where it should be and explain why you still can't see it. Explain completely the process you used in deducing their locations using the technique described in Box 2-2.

Example

A star which is 3 hand widths above the horizon in the north-west has an altitude of 30 degrees (3 hand widths times 10°) and is 4 and 1/2 hand widths (45°) around from due west (270°), so it has an azimuth of 315°.

Tricks of the Trade

1) The stars in Table 1 are the same as in Project 1-2, so you can copy their names, coordinates and visual magnitudes.

2) Start with the stars in the west first then work eastward.

3) *Be sure to hold your star map correctly oriented to the horizon, canted at an angle with the north end of the map pointed towards the north celestial pole.*

4) *Let your star atlas help you use stars you know to find stars you don't know. If two stars point at a third on your star map, they will do so in the sky. If two stars are 30° apart on the star map, they will be 3 hands apart in the sky.*

4) *Compare the altitudes and azimuths of stars in the same constellation to see if your observations make sense. For example, if both Rigel and Betelgeuse are in the South, the azimuth of Rigel will always be greater than the azimuth of Betelgeuse (if observed at the same time), while Betelgeuse will always have the greater altitude.*

Questions

1) What was the most common star color? Explain.

2) Which stars were below the horizon? Explain the process you used to determine that they were below the horizon.

3) Explain why starting with the stars in the west and working eastward is a good idea.

Table 1: Bright Stars

Common Name	Bayer-Flamsteed Name	RA H	M	Dec °	'	App. Visual Mag	MAP #	Alt °	Azi °	Color	Time
	α And										
		1	38	-57	14						
	α UMi										
		2	08	23	32						
	α Tau										
		5	14	-8	12						
	α Aur										
		5	26	6	21						
	β Tau										
		5	55	7	24						
	α Car										

Table 1 Con't: Bright Stars

Common Name	Bayer-Flamsteed Name	RA		Dec		App. Visual Mag	MAP #	Alt	Azi	Color	Time
		H	M	°	'			°	°		
		6	45	-16	42						
	ε Cma										
		7	40	5	14						
	β Gem										
		10	09	11	59						
	α Cru										
		12	47	-59	41						
	ζ UMa										
		13	26	-11	10						
	β Cen										
		14	15	19	12						
	α Cen										
		16	29	-26	26						
	λ Sco										
		18	37	38	47						
	α Aql										
		20	42	45	17						
	α Psa										

Project 2-2
Diurnal Motions

Use the same procedure as in Project 2-1 to measure the altitude and azimuth of stars. Sidereal time tells you how far around the celestial equator the first point of Aries (the zero point or origin of the celestial coordinate right ascension) is from your local meridian. The easiest way to determine the sidereal time is to just identify a star on your meridian and note its right ascension. Since the right ascension of the star tells you how far around the celestial equator it is from the first point of Aries and it is also on your meridian, then its right ascension tells you the sidereal time.

1) Pick four stars and measure their positions in human coordinates each half hour. Record your observations in Tables 1, 2, 3 and 4, respectively, for each star. The *approximate* human coordinates for each star should be:

   North — 10° to 30° from Polaris
   East — Altitude below 30°, azimuth between 70° and 120°
   South — Altitude between 20° and 40°, azimuth between 150° and 210°
   West — Altitude between 20° and 40°, azimuth between 250° and 300°

   Using The Trained Eye Star Atlas, find and identify the stars. Record their names, and coordinates and magnitudes in Tables 1, 2, 3 and 4 as appropriate. Explain how you identified each star.

2) For each hour of *local solar time,* face south and look for a star on your meridian. Identify that star from The Trained Eye Star Atlas. Record its name, celestial coordinates, magnitude and *local solar time* in Table 5. The *local sidereal time* is equal to the right ascension of the star at each hour of *local solar time* listed in Table 5.

Tricks of the Trade

1) *Let your star atlas help you use stars you know to find stars you don't know. If two stars point at a third on your star map, they will do so in the sky. If two stars are 30° apart on the star map, they will be 3 hands apart in the sky.*

2) *Pick stars that are brighter than 3rd magnitude. Avoid faint stars.*

3) *Take extreme care in measuring the positions of each star. Repeat your measurement a couple of times.*

4) Take extreme care that you have the right star before taking a measurement.

Questions

1) Using the data from Tables 1, 2, 3 and 4, describe how each star moved.

   Hint: Make a plot of altitude vs azimuth for each star to show its motion. Backtracking a star's motion you can predict where and when it rose. By tracing a star's motion forwards in time, you can predict where and when it will set.

2) Predict the azimuth and time the star rose or the azimuth and time you expect it to set.

   Hint: If its in the east, predict the time it rose. If its in the west, predict the time it will set.

3) How fast did each star move (°/hr)? How fast in altitude? How fast in azimuth? Which star moved slowest? Which moved fastest?

   Hint: Measure the total distance moved in millimeters and convert to degrees using the degree scales on the side of the plots.

4) Where in the sky do you expect stars to move the slowest? Where do you expect them to move the fastest? Explain why.

5) Using your local sidereal time for 9:00 pm, where (approximately) in human coordinates would you expect to find the following celestial coordinates? Explain how you deduced your answers.

   Hint: lean over and align your body with Polaris — put yourself into celestial coordinates. Then you should be able to sweep your hand around the celestial equator the correct number of hours from your meridian to see where the following celestial coordinates lie in human coordinates.

0°	Declination	6^h	Right Ascension
0°	Declination	12^h	Right Ascension
0°	Declination	18^h	Right Ascension
0°	Declination	0^h	Right Ascension

Table 1: Northern Star

Common Name	Bayer-Flamsteeed Name	RA h	m	Dec °	'	Alt °	Azi °	Color	Time

Table 2: Eastern Star

Common Name	Bayer-Flamsteeed Name	RA h	m	Dec °	'	Alt °	Azi °	Color	Time

Table 3: Southern Star

Common Name	Bayer-Flamsteeed Name	RA h	RA m	Dec °	Dec '	Alt °	Azi °	Color	Time

Table 4: Western Star

Common Name	Bayer-Flamsteeed Name	RA h	RA m	Dec °	Dec '	Alt °	Azi °	Color	Time

Table5: Local Sidereal Time

Common Name	Bayer-Flamsteed Name	RA h	RA m	Dec °	Dec '	Local Time	Sidereal Time

Chapter 3
Extending the Eye's Reach

The eyes have been called the windows of the soul, but for the heavens they provide only a peep-hole. Although marvelous in their abilities, eyes excel in daylight not darkness. Eyes see only shadows on the Moon, not its mare, mountains and craters. Except for a few glorious exceptions, stars appear as flickering pinpoints of light, bereft of color. Our galaxy, a spiraling collage of two hundred billion suns, presents only a meandering band of pale light easily lost in the glare of incandescence from used car lots. To understand the universe in detail, we must first see the detail — detail that is invisible to the unaided eye. This thought is not peculiar to our time and place. Galileo Galilei, an Italian astronomer, physicist, chemist, mathematician, entrepreneur of the late Renaissance, understood this as did many of his contemporaries.

In 1609 the telescope arrived at Galileo's home town of Venice in the hands of a traveling salesman from the Netherlands. We don't know which Dutchman to credit as its inventor. In truth the technology to build telescopes was commonplace. Telescopes had, no doubt, been invented and forgotten many times before. We shall credit Galileo as the true discoverer, for he discovered its cosmic potential to extend our eyes and minds into the universe.

What is this marvelous instrument that Galileo discovered? What is a telescope? More than what's advertised, as in the "typical" shown in Figure 3-1. What is a telescope? Something more than this glib recitation of active verbs, positive adjectives and strong nouns. What does power mean? It sounds good and strong. How about focal length or detail? What is a telescope? Complete Project 3-1 and you'll know.

ASTRONOMICAL SPECIAL

Feature packed 600 power telescope puts the stars in your eyes

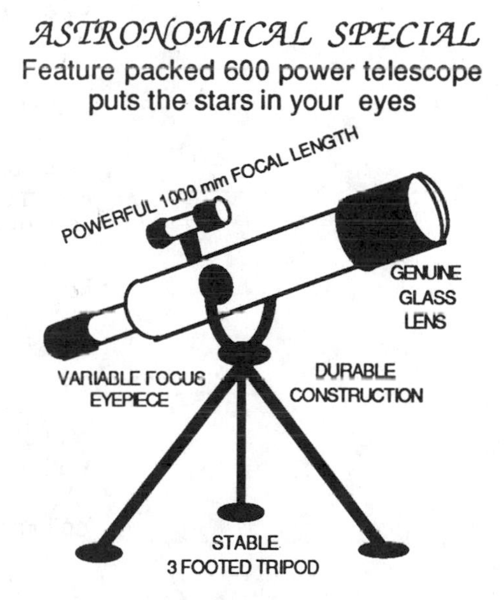

1000 mm focal length. Twice as long for twice the observing power.
600 power magnifies the smallest detail to spectacular proportions.
Variable focus eyepiece. Full range focusing from a soft blur to crystal clarity.
Sturdy cast iron construction. Rustproofed to survive the roughest observing conditions.
Three footed tripod features double lubricated ball bearings for smooth tracking.
Finest optical grade crown glass singlet objective lens. Multi-polished for sharper images.
Two telescopes for the price of one. Includes a bonus telescope to show the small scale structure of the universe.
Full color instruction manual and guide to the universe.

Figure 3-1

A Telescope Is What It Does

Who would have imagined that lenses, those rounded lumps of glass, could produce such a spectacular synergism? A simple pairing of two lenses compensates for the eye's weakness by letting you see fainter and further. However, as you saw in Project 3-1, the telescope's defects are as obvious as its benefits; you see fainter, further, blurry and distorted. Galileo, with his well trained eye, could overlook the defects in his crudely ground lenses. Many of his contemporaries could not. They could not distinguish between the divine truths Galileo's telescope revealed and the aberrations it created.

Nowadays we think of telescope in the plural to acknowledge the plurality of telescope designs created in response to the promise of what Galileo saw in his, when first he turned it to the heavens. A lot of work by a lot of people to improve the telescope — improvements in materials, fabrication and optical components — to make the toy into a scientific instrument.

The Objective

Before getting into the two types of objectives that define the two main classes of telescope designs, let's review the objective's attributes; what it does in a telescope. The objective determines two properties of a telescope uniquely: — light gathering power and resolution — and the two in collaboration with the eyepiece — magnification and field of view (discussed when you get to eyepieces). Light-gathering power often makes more sense when described in terms of *limiting visual magnitude*, i.e., how faint you can see (Table 3-1). Your unaided eye can see stars brighter than sixth magnitude i.e., a limiting visual magnitude of 6. A telescope, because of its larger objective lens diameter, has greater light-gathering power and aids your eye to see a fainter (higher) limiting visual magnitude. Using a familiar equation from high school geometry that relates the area of a circle to its diameter, you can contrast the light-gathering power of your unaided eye against that of an 8-inch diameter objective (typical of many amateur telescopes) by comparing their collecting areas:

$$\textit{Collecting Area} = \pi \left(\frac{\text{diameter}}{2}\right)^2$$

The pupil of your eye has, upon full dark adaptation, a diameter of about 6.3 mm (1/4th of an inch) for a collecting area of 31.7 mm^2. An 8-inch diameter objective (196 mm) has a collecting area of about 32,400 mm^2. These meaningless numbers produce a meaningful one when you take the ratio of the two collecting areas i.e., divide 31.7 into 32,400 to get 1024. This ratio means an 8-inch telescope has a collecting area 1024 times bigger than your eye and aids your eye to see 1024 times fainter. How do you express this 1024 times fainter in terms of limiting visual magnitude? A ratio of 1024 times corresponds to a magnitude difference of 7.5 (Chapter 1). Add 7.5 to the limiting magnitude of your unaided eye (6) and you get the limiting visual magnitude for your 8-inch diameter telescope aided eye: 13.5. If you want to *see* even fainter (the astronomer's holy grail), get a even bigger objective. What limits the size of the objective? Usually your wallet. Bigger not only is better, it's also more expensive. Terrestrial viewing, in contrast to astronomical viewing, generally has a surfeit of light, and a small objective suffices.

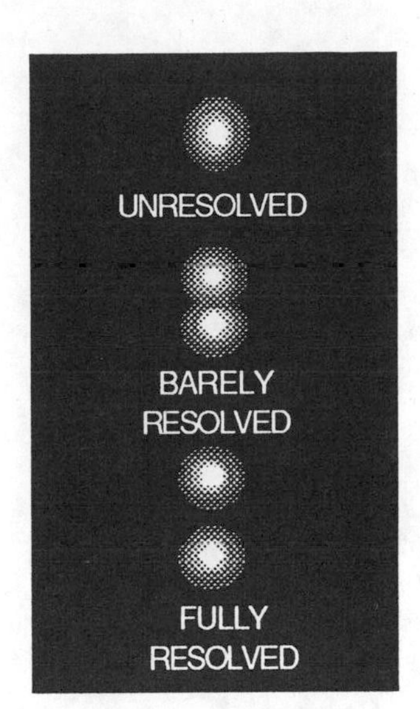

Figure 3-1

The last column in Table 3-1 relates the diameter of an objective to its *resolution*; its ability to show two stars as two distinct objects instead of one (think of how a distant car's two headlights appear as one). Objectives focus stars into small blurred disks not into the pinpoints of light they seem. The fault, dear Brutus, lies not in the stars nor in objectives but in the wave nature of light; no amount of lens polishing can improve it. Increasing objective diameter, however, can: the bigger the objective, the smaller the blur disks and the closer two star images can be and still be seen as two (Figure 3-2). You see this in Table 3-1: as objective diameter goes up, theoretical resolution (the closeness at which two stars are recognizable as two) goes down.

It's often a judgement call, but if you can see the waist in their merged images ("Barely resolved" in Figure 3-2) then you know there are two stars, not one. You can calculate theoretical resolutions for any objective using the following equation for visual light:

Table 3-1: Properties of the Objective

Objective Diameter mm	Objective Diameter inches	Light Gathering Power	Limiting Visual Mag.	Theoretical Resolution "
6.35	0.25	1	6.0	21.73
25.40	1.00	16	9.0	5.43
50.80	2.00	64	10.5	2.71
152.40	6.00	144	11.4	0.90
203.20	8.00	1024	13.5	0.67
254.00	10.00	1600	14.0	0.54
406.40	16.00	4096	15.0	0.34
812.40	32.00	16384	16.5	0.17
1624.40	64.00	65536	18.0	0.08
3248.80	128.00	262144	19.5	0.04

$$\text{Resolution} = \frac{138\,''}{\text{objective diam in millimeters}}$$

or

$$\text{Resolution} = \frac{5.4\,''}{\text{objective diameter in inches}}$$

Resolution is measured in ", otherwise known as arc seconds or 1/3600th of a degree. The wavelength of light also affects resolution, and a more quantitative form of the equation will include the wavelength dependency. From the theoretical resolutions listed in Table 3-1, you can see that, just as for limiting visual magnitude, the bigger the objective the better the resolution.

Aberrant Behavior

In reality, both limiting visual magnitude and resolution are limited by environmental effects: light pollution and atmospheric blurring. Light pollution doesn't affect your health, only your sight. Just think about how much street lights, neon signs and used car lots emit. Most of their light emissions are wasted, either bouncing off of the ground back up into the air or aimed up to start with (like searchlights). Although most of this wasted light shines back into space, the atmosphere scatters a fair percentage back into your eyes. In a city like Los Angeles, the sky glows so brightly at night that you can read by it — well, headlines at least. The sky glows with pollution and the stars fade into invisibility. The visual magnitude at which nature loses out to man is your limiting magnitude. The moral? It's hard to see stars with light bulbs shining in your eyes. Now what about atmospheric blurring?

Look down a long road on a hot day and the very air shimmers. It may be less noticeable straight overhead, but the atmosphere shimmers in every direction. This atmospheric turbulence blurs stars into small disks in an effect called *seeing*. The more turbulent the atmosphere, the bigger the blur disks, the worse the seeing. As long as this blurring is less than the theoretical resolution of your objective, you don't notice it, so who cares? The trouble is, seeing never gets better than about 0.5", which means the resolution for all telescopes 15 inches in diameter or bigger is dominated by seeing and except for a few that use flexible mirrors and computers to compensate for the blurring of the atmosphere, none achieve their theoretical capabilities (Table 3-1).

Despite this, astronomers are ecstatic with 0.5" seeing — it's the best you can get from the Earth. However , it takes an extremely good observing site with extremely still air and extremely good weather conditions. Usually seeing varies from 1" on a good night to 20" on a lousy night, with typical values around 2-5". Thus, on most nights, seeing dominates resolution for most sizes of objectives.

With careful choice of observing site (dark) and weather (clear and still), you should see the benefits of your telescope's objective. This is especially important since the objective, the first component of your telescope that starlight enters, fixes the quality of the image that the eye will see. A lousy objective will produce a lousy image that no eyepiece can make better. What makes an objective lousy besides lousy craftsmanship? The same intrinsic problems Galileo saw.

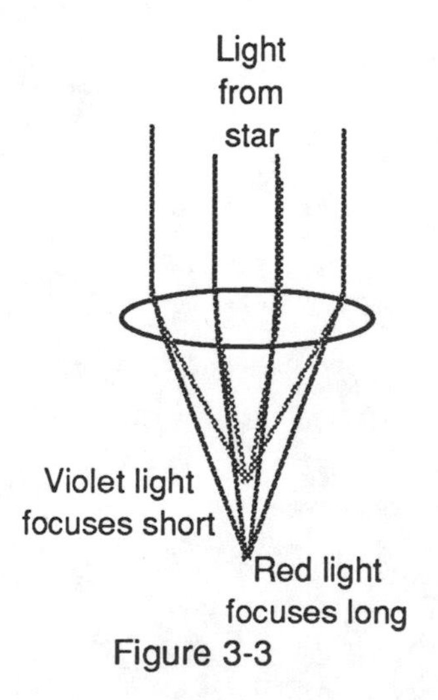

Figure 3-3

Galileo saw the same faults in his objectives that you saw in yours (Project 3-1): *chromatic* and *spherical aberration.* Chromatic aberration produces colored haloes around objects in an image. Lenses focus light by refraction — the same process that a prism uses to bend sunlight into a rainbow of colors. Just like a prism a lens refracts different colors of light by different amounts (Figure 3-3). The bending is necessary to produce an image but each color of light, bent by a different amount, focuses at a different focal length. Focus the red, and blue goes out of focus (you see blue haloes). Focus the blue, and red goes out of focus (you see red haloes).

Even if you could eliminate chromatic aberration, your objective lens will still produce spherical aberration. Most lenses have spherical curvatures and the outer part of the lens has a different focus than the inner parts of the lens (Figure 3-4). The result? A blurred image. You can reduce spherical aberration by using a long focal length objective: the longer the focal length, the less the curvature of the lens, the less its spherical aberration.

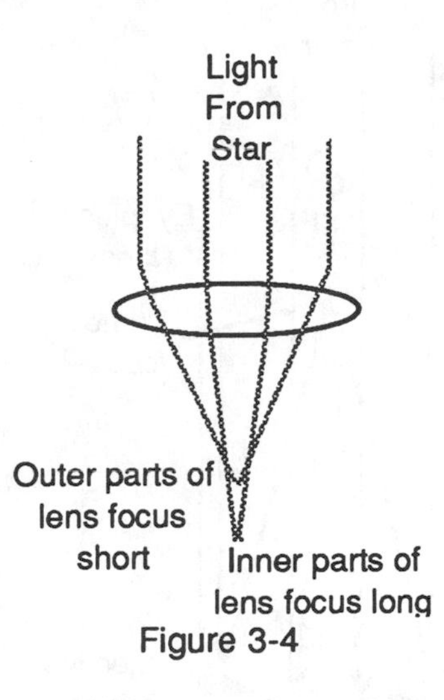

Figure 3-4

What's good for spherical aberration is good for chromatic aberration — an adage (with some truth) that early telescope makers carried to extreme. In their attempts to reduce aberrations some astronomers developed telescopes over 200 ft. long (yes, that's what it took to ameliorate spherical and chromatic aberration). Imagine pointing such a telescope in any kind of breeze! There has to be a better way.

There are, in fact, two better ways; the *achromatic* (without color) *doublet* lens and the parabolic mirror. The doublet objective lens works by pairing a *convex* lens of one type of glass with a *concave* lens of a different type of glass. With proper matching of the two lenses, their chromatic aberrations cancel each other out (almost) and the light still comes to a focus. Mirrors image by reflection, not refraction, and so avoid chromatic aberration completely. Using a concave mirror with a parabolic curvature instead of a spherical curvature completely eliminates spherical aberration (but does create another problem, coma, which we will ignore for now). Oh, by the way, what are convex and concave?

A convex lens or mirror is thicker in the middle than at the edges, a concave lens or mirror is thicker at the edges than in the middle. You probably have a few examples of each around the house. A magnifying glass uses a convex lens. If you're nearsighted, your eyeglasses use concave lenses, makeup mirrors use concave mirrors, and the right side rear-view mirror on your car, the one that says "objects are closer than they look", uses a convex mirror.

The dichotomy in choices, between using a lens or a mirror for the objective, naturally divides telescopes into two categories: *refractors* (which use objective lenses) and *reflectors* (which use objective mirrors).

Refractors

You built a simple refractor in Project 3-1 using a singlet lens for the objective (Figure 3-5). Building a good refractor takes a lot more work, lenses, and money. Most modern refractors use long focal length doublet lenses to minimize chromatic and spherical aberration. However, long focal length means small field of view ,so what do you do if you want wide field of view? As you know (Project 3-1) you need a short focal length objective, but a short focal length doublet can't minimize chromatic and spherical aberration as well as a long focal length doublet (despite some manufacturer's claims to the contrary).

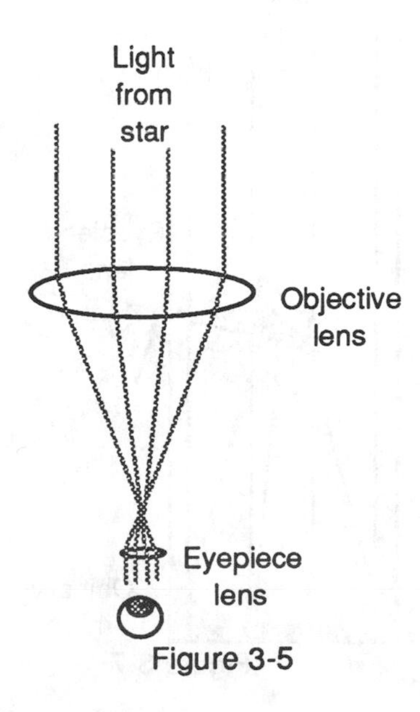

Figure 3-5

Designing a refractor requires a tradeoff between benefits (such as field of view) and detriments (such as chromatic and spherical aberrations). Use more lenses in the objective (say three for a *triplet*) or some other more exotic lens combination and a manufacturer can produce an excellent wide-field refractor and empty your wallet faster. Which brings us to our consumer advisory. No matter how many lenses a manufacturer uses, no matter how many anti-reflection coatings, no matter how much the promotional literature claims, believe your eyes. Focus in on a bright blue star (like Vega) — if at its sharpest focus you still see a blue halo (or a red halo), you have a paperweight not a refractor.

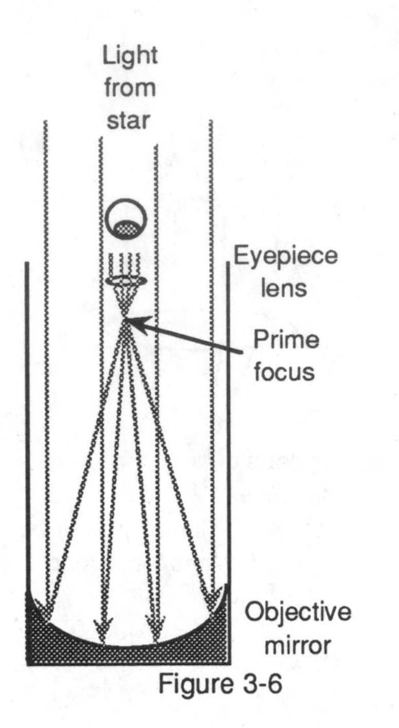

Figure 3-6

Reflectors

Sir Isaac Newton, the sage of gravity and optics, devised the first workable reflecting telescope and gave his name to the design: the *Newtonian*. Although he knew parabolic mirrors to have zero spherical and chromatic aberration, he recognized their one big headache: the image lies in front of the mirror (at the *prime focus*). To see the image you have to stick your head in front of the mirror ,which blocks the light (Figure 3-6). For some of the largest telescopes, like the 200 inch diameter Hale telescope at Mount Palomar, your head — in fact your whole body — is much smaller than the objective's diameter, and you can ride at the prime focus and still see the stars.

Newton's 3-inch-diameter objective was a bit small for this solution. So instead of his head, he put a tiny second mirror (the *secondary*) in the converging beam of the objective mirror (the *primary*). The secondary, a small flat mirror tilted at 45 °, reflected the converging beam to a focus just out the side of the telescope tube where Newton could see it without his head getting in the way (Figure 3-7).

The Newtonian reflecting telescope remains the most popular type for amateur astronomers owing to its low cost compared to refractors and more complex reflector variations. There are indeed many other versions of reflectors, each with a different combination of primary and secondaries, different curvatures and some combining refractive elements. As always, the more the complexity the more the cost.

Objective Craftsmanship

Beyond the inherent advantages and disadvantages of their design, both refractors and reflectors are only as good as the craftmanship that goes into their making. Objective mirrors must have smooth surfaces to avoid the fun-house mirror effect. *Flatness* (a strange term to apply to curved surfaces) measures how well the actual curvature of a mirror matches the desired theoretical curvature. If it's high here and low there, it isn't very flat. Flatness is typically quoted in fractions of the wavelength of visible light (around 0.0005 mm); 1/4th wave is minimum quality, 1/8th wave is more typical, 1/20th wave is superior. Obviously we're not talking about very tall hills or deep valleys when we talk about flatness.

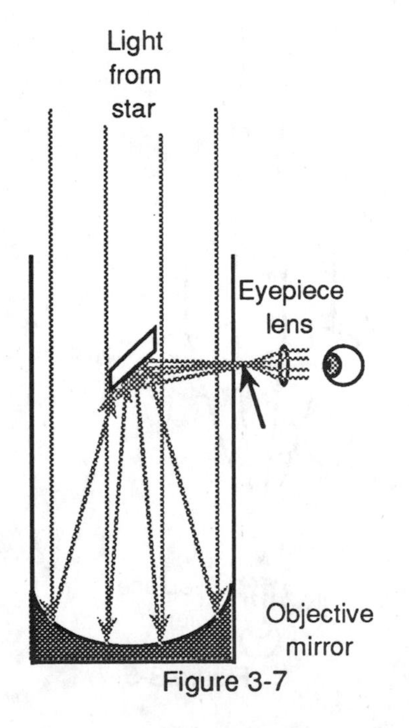

Figure 3-7

Lenses must be of equal flatness for equal reasons. Doublet objectives (and triplet and quadruplet and etc.) must also have each lens well matched with the others to eliminate chromatic aberration. Because it takes more lenses to make a refractor than mirrors to make a reflector and each lens has two surfaces that must be perfect compared to one surface for a mirror, refractors typically cost three to four times more for the same aperture. Plus you have the difficulty of making large lenses that won't sag under their own weight (distorting their flatness). Reflectors don't have this problem since you can put support brackets all over their backsides. Some people claim advantages to refractors over reflectors, but cost is certainly not one of them. Again, the ultimate test of any telescope (refractor or reflector) is what you see through it.

The Eyepiece

Most large telescopes operate as large cameras with a piece of photographic film or other electronic detectors placed at the focus of the objective to capture the image and as such they don't need eyepieces. Remember that the objective determines two properties of a telescope uniquely: light-gathering power and resolution. Seeing through a telescope requires an eyepiece which, in addition to letting you see the image, helps define how big the image looks (magnification) and how much of it you see (field of view) in collaboration with the objective (Project 3-1). Since the eyepiece comes between the image and your eye, you need a good eyepiece to see a good image. So say hello to your old friends chromatic and spherical aberration, and your new friends, *eye relief*, *field of view*, and *field flatness* (only a partial list). Just like for the objective, you can use combinations of lenses to minimize chromatic and spherical aberration in an eyepiece.

Eye relief, as you recall from Project 3-1, is how close to the eyepiece to hold your eye. If your eyepiece uses only a single convex lens, eye relief equals the focal length of the eyepiece lens. No problem for a 50 mm focal length lens, but how about a 10 mm, or a 5 mm focal length? Looking through a 5 mm focal length lens can be as much fun as sticking a finger in your eye. With clever design, however, you can move the eye relief of a 5 mm lens more than 5 mm out from the lens, making it easier (and less painful) to use.

Field of view involves both the objective and the eyepiece. There's not much you can do to change field of view for an objective short of buying another telescope. Remember, you measured field of view in Project 3-1 for simple (single) eyepiece lenses, not combinations of lenses. Different eyepiece designs of identical focal length can have significantly different fields of view.

Field flatness. If you wear glasses, you know the world looks a bit distorted. The thicker your lenses, the worse the distortion. Squares aren't square, they either bow in or bow out — they don't look flat. Eyepieces have this same problem. What's worse, the focus may vary across the field; stars near the edge are out of focus, while stars in the center are in focus. As a rule, the wider you try to make the field of view, the worse the field flatness.

Eyepiece designers trade off all five of these "features" plus trying to keep the number of lenses in the eyepiece to a minimum. Why limit the number of lenses? The more lenses, the more light lost in getting through them. Even perfectly clear lenses absorb or reflect some light (typically 4% at each surface). This may not seem like much light loss until you remember that each lens has two surfaces, so each lens costs you 8% of the light from reflection alone. Multiply this by three or four lenses, and you've lost a lot of light. Of course, the light isn't really lost, it just scatters all over the inside of the eyepiece. You eventually see it as ghost images (bad enough), diffuse glow (just as bad), or worse. Special anti-reflection coatings applied to the lens surfaces reduce reflection losses to below 1% per surface. This ameliorates, but doesn't eliminate, the scattered light problems. The ultimate limit on the number of lenses in an eyepiece is cost. The more you put in the better (potentially) the image and the greater (definitely) the costs. So how do you tell which eyepiece is right for you? As with objectives, seeing is believing — buy the best seeing eyepieces you can afford.

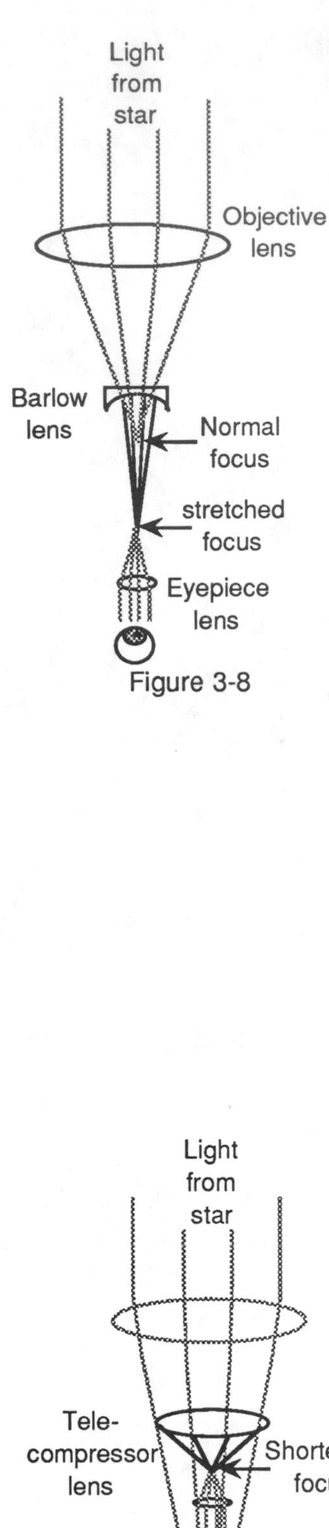

Figure 3-8

Objective Versus Eyepiece

The objective is the heart of your telescope, for it establishes the principal capabilities of your telescope. Resolution is defined by the objective. Limiting visual magnitude is defined by the objective. Cost is defined by the objective. Magnification and field of view are limited by the objective. True, you can always change eyepieces to tailor the magnification and field of view of your telescope — of your objective — within reason. So what's reasonable? Before getting into this subject, let's redefine the dimensionless version of focal length: f-ratio. *F-ratio*, as you remember from Project 3-1, is the dimensionless (no units) number you get when you divide the focal length of an objective by its diameter. For example, an 8-inch objective with a focal length of 64 inches has an f-ratio of 8 (expressed as f-8); the focal length is eight times as long as the objective is wide. Notice that it doesn't matter whether you use inches or mm to calculate f-ratio, you get the same answer.

If you want a wide-field (or rich field) telescope, you need a short f-ratio objective (f-4 to f-5). If you want high magnification, you need a long f-ratio objective (f-10 and above). If you want something in between, you need an f-ratio in between (f-6 to f-9). Why does this seem to refute what you just learned in Project 3-1 — that, regardless of the objective, you can get any magnification or field of view you want with the right eyepiece? True in theory, impossible in practice. To get high magnification out of a wide-field objective, you need extremely short focal length eyepieces (< 7 mm). These eyepieces are a pain to make, a pain to use, and a pain in the wallet.

Well if you can't get high magnification out of a short f-ratio objective, can you get a wide field out of a long f-ratio objective? Here again you run into the limitations of eyepiece construction. The longest focal length eyepieces generally available are around 40 mm — still not long enough to convert your long f-ratio objective into a wide-field instrument. You again find problems with eye relief, construction and cost when you push beyond this upper limit on eyepieces.

So the objective's f-ratio defines what type of telescope you have, wide-field, low-magnification or narrow-field high-magnification. Eyepieces modify field of view and magnification within the useful range set by the objective. You can work around this limitation (or feature) by using one of two optical accessories — but at a cost in image quality and dollars. The first of these optical devices is called a *Barlow lens.*

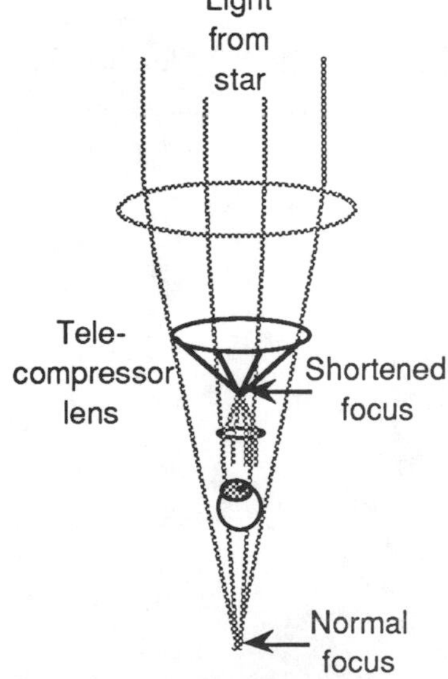

Figure 3-9

A Barlow consists of a concave reducing lens (or combination of lenses) that changes the *effective focal length* of the objective, essentially converting a short f-ratio into a long f-ratio (Figure 3-8). Typical Barlows will double or triple the effective f-ratio of your objective, thus doubling or tripling the magnification of your telescope. That's the good news. What is the bad news? The Barlow introduces another optical element (more light loss, more aberrations), and restricts the field of view.

Going the other way, there are devices called *tele-compressors* which shorten *effective focal length* (Figure 3-9). They too, just like Barlows, introduce more optical elements, light losses, and aberrations.

Given their side-effects, the best solution remains to pick the objective for your telescope that fits your desires for light-gathering power (diameter), resolution (diameter), field of view (f-ratio) and magnification (f-ratio).

Why A Finder?

Regardless of your choice of objective and selection of eyepieces for your telescope there is a special accessory you can't do without: a small telescope bolted onto your big one. After all, that's what makes it an astronomical telescope, isn't it? Many manufacturers use this popular perception to their advantage, but what they bolt on may be worse than useless.

This small telescope serves a useful purpose implied by its name: *finder telescope* or *finder* for short. A finder helps you *find* stars for your *main telescope* (the one the finder's bolted onto). A finder works for your main telescope just like a telescopic gun sight does for a rifle; the gun sight helps you aim the gun, the finder helps you aim the main telescope. A finder, just like a telescopic gun sight, has a cross hair that you must pre-align to the main telescope before you can aim it properly. During the daytime, point your main telescope at some far distant object and adjust the finder until its cross hairs cross on the same object. After that, whatever you see on the finder's cross hair is what you'll see through the main telescope (unless you bump it). Still, why do you need a finder? After all, isn't your main telescope a telescopic sight?

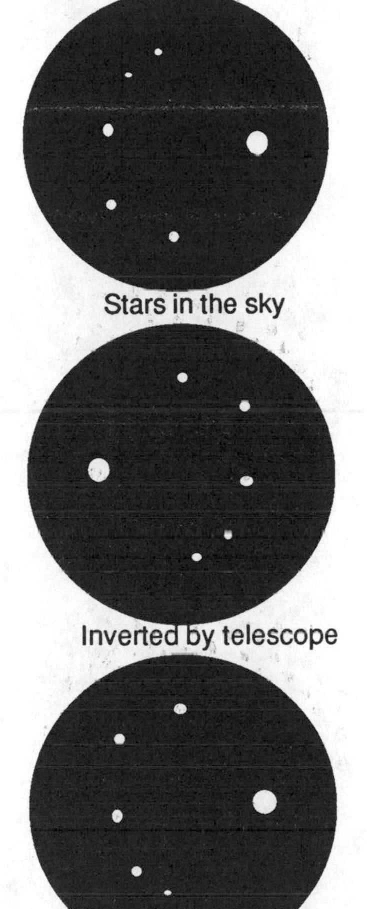

Figure 3-10

At best, your main telescope, regardless of its f-ratio, has a field of view 1° or so — pretty small compared to the unaided eye's 180°, and pretty hard to aim at a star. A good finder bridges the field-of-view gap to place what you see with your unaided eye into the main telescope. Some finders (and main telescopes) use a *star diagonal*, an attachment that lets you look sideways into the finder rather than straight through. The mirror or prism used in the star diagonal introduces a mirror flip into the image (Figure 3-10). Not only do things look upside down, they also look backwards. Upside down you can deal with by just turning the star map upside down but what about backwards? There you'll have to learn to flip the map back over in your mind.

A good finder should have intermediate field of view (around 5°), mild magnification (6 to 10 times), good light gathering power (limiting visual magnitude about 10th, which implies a minimum acceptable aperture of 2 inches / 50 mm). Ultimately, since *finder telescopes* are refractors they require the same good optical design as any refractor (achromatic doublet objective, quality eyepieces, etc.). A fact too often lost in the merchandising of inexpensive (i.e. cheap) telescopes.

Box 3-1 Perception and Optics

Your eyes too have their own aberrations. For example, when your pupils fully dilate, the spherical aberration in your eye lens is at its worst and the image on your retina blurs. That's why people look better under candlelight — you don't see the wrinkles.

If you wear eye glasses, you have become used to how they distort the world around you. It is only when you get a new prescription that you are reminded how they distort reality (until your mind becomes adapts to your new prescription). Off to the periphery of your vision, you'll see shapes distorted — straight lines curve, objects move funny.

It takes a few days, but your trained eye acclimates to the new lenses, and you don't see their distortions. When looking through a telescope ,you use the same skills to learn which parts of an image are real and which parts are due to the optics. That's why Galileo could see the stars through the aberrations of his telescope. He trained himself on what was real and what was the side effects of his telescope.

When looking through a telescope, don't squash your glasses up against your face trying to get close enough to the eyepiece to see something. Take your glasses off and adjust the telescope for your eyes and you'll see everything those with perfect vision can.

Project 3-1
Lenses and Telescopes

Now's your chance to learn the the names and purposes of the optical parts of a telescope and the basic characteristics of a telescope:

- Light-Gathering Power
- Magnification
- Field of View
- Resolution

and the common defects in a telescope:

- Chromatic Aberration
- Spherical Aberration

The type of telescope you will construct is a combination of simple lenses (that is, single lenses like those in eyeglasses). All of the basic characteristics of a telescope (listed above) are derived from the two dimensions which are associated with all simple lenses: diameter and focal length.

In your equipment box, you'll find the following;

1 Optical Bench
4 Optical Bench Clamps
3 Optical Bench Lens Holders
1 Adjustable Diaphragm
4 Convex Lenses (nominally 5, 10, 20 and 30 cm focal lengths)
1 White Cardboard
1 Frosted Glass
1 Flexible Plastic Ruler
1 High-Intensity Lamp (for local illumination)

At the front of the room, you'll see several targets:

1) Three light bulbs in a triangular constellation, each bulb marked with crossed arrows.

2) A line filament light bulb covered with blue plastic from top to bottom and red plastic on the bottom half so that it transmits blue light through the top half and red light through the bottom half.

Note: The dark blue filter transmits some red light (the "red leak"). This is used to advantage to attenuate the red image so that when the red filter is added to the bottom of the bulb, the blue image from the top half and the red image from the bottom half are of comparable intensity.

3) A foil target with a pair of closely spaced pinholes illuminated from behind through a red filter.

4) The numbers 1 to 16 spaced 6 inches apart written left to right on the blackboard and written upside down from right to left just under the upright version.

You will now go through a graduated series of experiments to illuminate the properties of simple convex lenses and how these properties combine to produce a telescope. In each experiment, you will be asked to make certain measurements and observations, record your measurements, analyze your data to deduce properties of the lens combinations you are working with and answer questions on those properties. At the end of the project, you will find questions on each of the experiments. You'll find it to your advantage to review the questions as you go through each experiment. So set your optical bench on the table and aim it at the three light targets.

1 2 3 4 5 6 7 8 9 10 11 12 13 14 15 16

Number line on blackboard

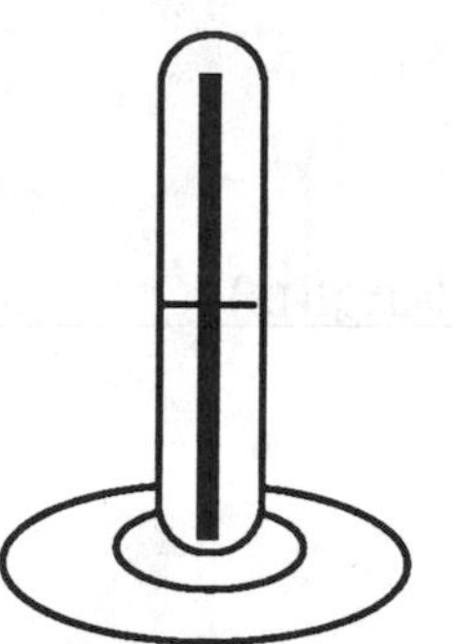

Line filament blub covered with blue plastic from top to bottom and red plastic on bottom half.

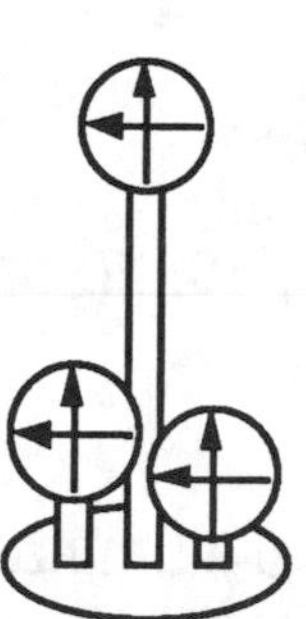

Three target light constellation

Red light covered with foil with pair of closely spaced pinholes

Part A: Simple Lenses

Diameter

Use the flexible ruler to measure the *diameters* of the four lenses in your equipment box. The lenses have nominal focal lengths of 5, 10, 20 and 30 cm. These numbers will be used to identify them. First record the diameters for each lens in Table 1.

Focal Length

The process for determining the *focal length* of the lenses is a little more involved:

1) Put the optical bench on your table so that the end where the ruler reads zero is nearest the back of the room. The other end should point at the lights on the table in the front of the room.

2) Put the white cardboard into a holder on the optical bench and slide it to position 0.

3) Put lens 5 into another holder on the optical bench near position 10.

4) Line up the lens and cardboard so that the three-light constellation at the front of the room shines through the lens onto the cardboard.

5) Slide the lens back towards 0 just until the three-light constellation *focuses* on the cardboard.

6) Read the distance between the cardboard and the lens off of the ruler on the side of the optical bench. That is the *focal length;* it is just the distance between the lens and the cardboard.

7) Have your partners each repeat steps 5 and 6 and enter your group average focal length for lens 5 in Table 1.

8) Repeat these steps for lenses 10, 20 and 30.

f-Ratio

The *f-ratio* is a dimensionless version of focal length. Dimensionless because it has no units. You may be familiar with *f-ratio* from photography. You can calculate the *f-ratio* for each lens by dividing the lens focal length by its diameter:

$$f\text{-ratio} = \frac{\text{focal length}}{\text{lens diameter}}$$

Table 1

Lens	5	10	20	30
Diameter				
Focal Length				
f-Ratio				

Use your data in Table 1 to answer questions 1 and 2.

Objective and Image Size

Now you are ready to dedicate one of your lenses to a special use, to use it as an *objective lens.* If you want to see a distant object (such as a mountain) better, the first solution most of us have is to hop in the car and make the distant mountain close. But there are cases in which this approach is not always practical, such as heavenly bodies. The solution is that of Mohamet: if you can't go to the mountain, then bring the mountain to you. This is exactly what the *objective lens* of a telescope does. It brings the distant *object* close by creating an *image* of it that you can get your hands on. Let's build an objective lens:

1) Put lens 30 into the holder just in front of the cardboard. Slide it forward until the 3-light constellation on the table at the front of the room is focused on the cardboard. This little picture you see on the cardboard is called the *image.* The objective lens has done its work, it has brought an image of the mountain close to you.

2) Measure the *image scale* by measuring the distance between two of the three lights in the target (use the pair that are farthest apart from each other). If your image looks small, then the image scale you measure will be small. If the image looks large, then the image scale will be large. Use the transparent ruler and record your value in the space provided in Table 2.

3) Repeat this for the other three lenses.

Table 2

Lens	5	10	20	30
Image scale				

Use your data from Tables 1 and 2 to answer questions 3 and 4.

Eyepiece Lens and the Real Image

Restore lens 30 to its proper place as the objective lens and focus it. Examine the image. Look for the arrows on the light bulbs. Move your eyes closer to see the fine detail. That's it, closer, closer... . Why does moving your eyes closer fail to work as you get close?

There is an easier solution which you may have hit on already. Use a magnifying glass to examine the image. Hold lens 5 in your fingers and try it as a magnifier. This lens is now your *eyepiece* lens, since you hold it near your eye. Keep looking for those arrows. What problems do you now have with your head?

Replace the white cardboard with the frosted glass. Slide both holders forward in the optical bench until the frosted glass is at position 10 and the objective lens is near position 40 or so. Focus the objective lens again. Now place another holder behind the frosted glass near position 5 and put your eyepiece lens (lens 5) in it. Slide it back and forth until you see a magnified image through it. Notice how much easier it is now to examine the image.

Speaking of images, this image is called a *real image,* which means, it is really there whether you can see it or not. Otherwise, you would not be able to touch it. Give it a try — remove the frosted glass and its holder and place your hand there.

Since the image is really there, you should still be able to see it in the eyepiece lens. Look through the eyepiece lens and slide it back and forth until the image comes back into focus.

Congratulations! You now have a simple refractor telescope! It is called a *Keplerian* telescope after *Johann Kepler* whom you may or may not remember from somewhere else in astronomy. How many lenses does it take to make a telescope?

Questions

1) Does focal length depend upon lens diameter or how rounded it is (how thick the lens is in its middle compared to its edge)? How does focal length depend upon lens curvature? Describe.

   Hint: Do more rounded lenses lenses have longer or shorter focal lengths?

2) For a fixed focal length, would different diameter lenses have different f-ratios? Describe.

   Hint: Does a larger diameter lens have a longer or shorter f-ratio compared to a smaller diameter lens if both have the same focal length?

3) Which objective gives the largest image? Which objective lens gives the smallest image?

4) How does the image scale depend upon focal length?

   Hint: As the focal length gets bigger or smaller, does the scale of the image get bigger or smaller?

Part B - Telescopes

Light-Gathering Power

1) Put the adjustable diaphragm into a holder at position 50 in front of the objective lens and slide it up tight against the objective lens.

2) Align it so you can see the numbers on the blackboard through the telescope and through the diaphram.

3) Open it wide and look through the telescope at the blackboard behind the lights.

   Hint: the best place to put your eyes, called the eye relief, is at the point behind the eyepiece where the objective, as seen through the eyepiece, fills the eyepiece. The distance of the eye relief point behind the eyepiece is equal to the focal length of the eyepiece. That is, you should have your eye 5cm behind a 5 cm focal length eyepiece.

4) Watch the numbers on the blackboard and have a partner close down the diaphragm as you watch. Answer question 5 based on what you see.

Spherical Aberration

The center of a lens focuses to a slightly different focal length than the outer parts of a lens resulting in *spherical aberration.* By varying the size of the diaphragm you can see the effects of systematically eliminating contributions to *spherical aberration* from the outer parts of the lens.

1) Open the iris diaphragm all the way back up.

2) With your eye at the eye relief point, watch the numbers on the blackboard and have a partner close down the diaphragm halfway as you watch. Answer question 6 based on what you see.

Chromatic Aberration

If you've ever looked through a childs' telescope and seen colored edges to objects, then you've seen *chromatic aberration.* Chromatic aberration is inherent in lenses since they, just like a prism, use refraction to bend light. Since each color bends by a different amount, each color comes to a slightly different focus.

1) Open the diaphragm all the way and use lens 10 as the objective and use lens 5 as the eyepiece.

2) Point the telescope at the filament bulb and focus the telescope for the blue half of the filament.

   Hint: You can accommodate quite a range of telescope mis-focus by adjusting the focus of your eye — exactly what you don't want to do for this exercise! Keep your eye at the same focus by relaxing your eye as if you were focusing a long way away. Every time you adjust the telescope's focus,remember to relax your eye's focus.

You will see a red blur surrounding the blue filament since the blue filter "leaks" red light in addition to blue. Measure the combined focal lengths of lens 10 and 5 (the distance between the two) and record your answer in Table 3 under blue focal length. Mentally note how wide the red blur looks compared to the blue image.

3) Focus the telescope for the red half of the filament. The blue half should look out of focus, while it's red "leak" image will be sharp. Measure the sum of the focal lengths of lens 10 and 5 (the distance between the two lenses) and record your answer in Table 3 under red focal length.

4) Repeat steps 1 through 3 using lens 20 as the objective and lens 10 as the eyepiece. Which telescope looked like it had the worst blurring?

5) Calculate the chromatic aberration for both lens combinations as follows:

$$\text{Chromatic Aberration} = \frac{(\text{Red Focal Length - Blue Focal Length})}{\text{Eyepiece Focal Length}} * \frac{\text{Objective Lens Diameter}}{\text{Objective Focal Length}}$$

This equation produces a measure for the red blur when the blue image is focused (or vice versa). The lens combination with the smaller blur has the smaller chromatic aberration. Compare your calculation with your perception in step 4. Does your perception match your calculation?

Table 3

Objective Lens	10	20
Eyepiece Lens	5	10
Red Focal Length		
Blue Focal Length		
Chromatic Aberration		

Use your data from Table 3 to answer questions 7, 8, 9 and 10.

Questions

5) What happens to the overall *brightness* of the image of the blackboard (not the line or lights) as the diaphragm gets smaller? Don't worry about contrast or clarity or sharpness, just brightness.

6) What happens to the clarity and sharpness of the image of the blackboard as the diaphragm gets smaller? Don't worry about brightness, just clarity or sharpness.

7) Which lens combination had the greatest chromatic aberration?

8) How does chromatic aberration depend upon focal length?

   Hint: As focal lengths increase, does chromatic aberration get better or worse?

9) Why does it depend upon focal length in this way?

   Hint: Which focal length lenses have the greatest curvature and hence bend the light from the bulbs the most?

10) Which lens would you expect to have the most chromatic aberration? Which lens the least?

Part C: Telescope Magnification

Magnification is simply how much bigger (or smaller) does a telescope makes a distant object appear.

1) Remove the diaphragm. Point the telescope at the numbers on the blackboard. Now comes the tricky part. Look through the telescope with one eye while looking at the numbers with the other eye and raise or lower the eyepiece until the numbers viewed through the telescope line up with the numbers viewed with your unaided eye. Superimpose the image of the big numbers (big means, as seen with the telescopic eye, they look big) over the tiny numbers (tiny means as seen with the unaided eye, they look small) in your brain. The vertical line drawn through the number 10 will help you align the two images in your mind. Pick two big numbers, one unit apart as seen with the telescopic eye, and line one of them up with a tiny number seen with the unaided eye. What you'll see is shown in the following figure.

2) Take a quick mental snapshot and note what small number lines up with the other big number. Calculate the difference between the two unaided eye numbers. Divide this by the difference between the two telescopic eye numbers (which, you remember, is one). This gives the *magnification* of the telescope, telling you how many times bigger the telescope makes the object look. Magnification is also referred to as power (check your department store catalog sometime for that expression). Record your result in the appropriate place in Table 4 and repeat for all the other lens combinations listed in Table 4.

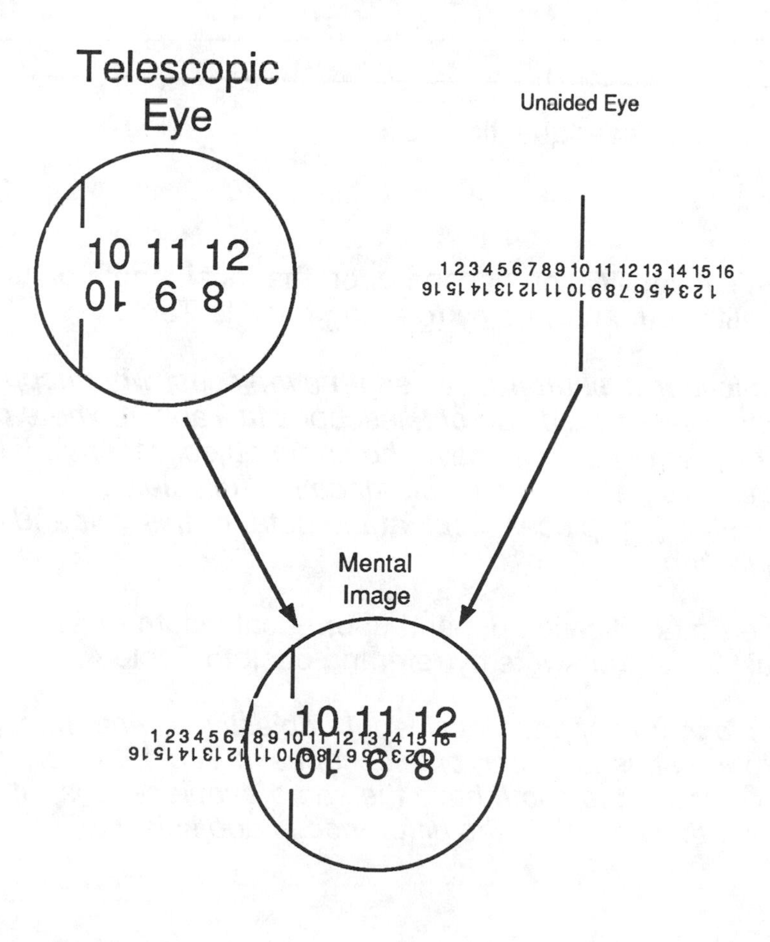

Table 4

Objective Lens Eyepiece Lens	30 5	20 5	20 10
Measured Magnification			

Use the data in Table 4 to answer questions 11 and 12.

Congratulations. You have now discovered the basic law of magnification;

$$\text{magnification} = \frac{\text{focal length of objective}}{\text{focal length of eyepiece}}$$

Use the focal lengths you measured (Table 1) to calculate magnification for the lens combinations listed in Table 5 and enter your results.

Table 5

Objective Lens Eyepiece Lens	30 5	20 5	20 10
Calculated Magnification			

Use the data in Tables 4 and 5 to answer question 13.

Questions

11) How does your telescope's magnification depend upon the focal length of its eyepiece? Explain and justify your answers by referring back to Table 4.

Hint: As you change from longer to shorter focal length eyepieces, what happens to magnification? Look at your results for a pair of telescopes in Table 4 where only the eyepiece lenses are different (they both have the same objective lens). If the magnification is different for this pair, then the difference is due solely to the difference in focal lengths of the eyepieces. Look at the data for this pair and give your impression to the question.

12) How does your telescope's magnification depend upon focal length of its objective? Explain and justify your answers by referring back to Table 4.

Hint: As you change from longer to shorter focal length objectives, what happens to magnification? Look at your results for a pair of telescopes in Table 4 where only the objective lenses are different (they both have the same eyepiece lens). If the magnification is different for this pair, then the difference is due solely to the difference in focal lengths of the objectives. Look at the data for this pair and give your impression to the question.

13) How do your measured magnifications (Table 4) compare to your calculated magnifications (Table 5)?

Part D - Telescope Field of View

Using the lens combinations listed in Table 6, record the distance between the leftmost and rightmost numbers you can see on the blackboard through the telescope for each lens combination. This is *field of view.* Record your results in Table 6.

Hint: Keep your eye at the eye relief point behind the eyepiece.

Table 6

Objective Lens	30	20	20
Eyepiece Lens	5	5	10
Field of View			

As you can see, some combinations of lenses allow you to see more and therefore have a wider field of view, while others let you see less and have a narrower field of view. Use the data in Table 6 to answer questions 14, 15 and 16. Remember to isolate the effects of the objective from the effects of the eyepiece: pick the pair with the same objective to see the effects of changing the eyepiece, pick the pair with the same eyepiece to see the effects of changing the objective.

Put lens 30 back in as the objective and lens 10 back in as the eyepiece. Put the diaphragm back in front of the objective and open it all the way up. Look through the telescope at the numbers (remembering to keep your eye at the eye relief point) and keep track of the leftmost and rightmost you can see while a partner closes down the diaphragm. Ignore the fact that it gets fainter or loses contrast. Use your observation to answer question 17.

Hint: Keep your eye at the eye relief point behind the eyepiece. You may find that as the diaphragm closes down you will have to move your head slightly to keep your eye at the eye relief point.

Questions

14) How does the field of view depend upon the focal length of the eyepiece? Justify your answers by referring back to Table 6.

15) How does the field of view depend upon the focal length of the objective? Justify your answers by referring back to Table 6.

16) Which combination of objective and eyepiece gave the largest field of view and which one (given your answers to the previous two questions), theoretically should? Justify your answers by referring back to Table 6.

17) Do you see more or fewer numbers as the diaphragm gets smaller? In other words, does the field of view depend at all upon the diameter of the objective lens? If so, how?

Part E - Telescope Resolution

This is the capability of a telescope to *resolve*, or show as more than one object something that is indeed more than one object. Do not confuse resolution with *spherical aberration*; the two arise from distinctly different causes.

1) Leave lens 30 in as the objective and use lens 5 as the eyepiece.

2) Start with the diaphragm two thirds closed (to minimize the effects of spherical aberration) and point the telescope at the pinhole target.

3) As your partner stops the diaphragm all the way down watch what happens to the size of the images of the pinholes. Use your observation to answer questions 18 and 19.

Questions

18) Do the sizes of the pinhole images change as the diaphragm gets smaller? Sketch what the images look like when you start and how they look as the diaphragm becomes very small.

19) As the diaphragm gets smaller, do the pinhole images get bigger and overlap so that they can't be seen individually, or do their images get smaller and become more easily separated? In other words, does the resolution depend at all upon the diameter of the objective lens? If so, how? Draw a sketch of what you see when the diaphragm is smallest.

Summary Questions

20) Compare your data for magnification and field of view (Tables 4 and 6) to determine if they are related. If they are related, how are they related? Describe.

Hint: As magnification gets larger, does field of view get larger or smaller? As magnification gets smaller, does field of view get larger or smaller?

21) How are light gathering power, spherical aberration and resolution related to each other and the diameter of the objective lens? Describe.

Hint: As light-gathering power gets larger, do spherical aberration and resolution get better or worse? Do all three improve as the objective diameter gets larger or smaller or do some get better and the others worse?

22) What effect does the eyepiece have on light-gathering power or resolution capability of the telescope? Should it have any effect? Explain.

23) Under what circumstances could two lights be resolved by the objective lens but still not appear as two images through the eyepiece lens? What could you do about it?

Hint: If something is too small to see, what do you do?

24) The more rounded (shorter f-ratio) a lens, the better or worse the spherical aberration and chromatic aberration? Explain.

Hint: A flat lens has no spherical or chromatic aberration.

25) Define the terms used in this project.

Chapter 4
Establishing a Firm Foundation

A telescope's mounting provides two valuable and contradictory services for your astronomical telescope: *pointing* and *steadying*. To reinforce their importance just grab your binoculars, step outside, and try to do the mounting's job yourself. Pick a bright star, say Procyon or Altair, and take a look. No doubt you find it hard just getting either star into view since binoculars reduce your field of view to only about 5° across. How little is 5°? Hold a hand up out to arm's length, make the international "OK" sign (thumb and first finger making a loop) and look at Procyon or Altair through that loop. That's 5°. No wonder you have so much trouble getting the right star into view. Even when you finally get Procyon or Altair to dance into view, it won't stop dancing. After all, your head is craned back in an awkward position and your arms quickly fatigue trying to steady your binoculars — to what avail? Even when you hold it perfectly steady ,the star still twitches to each pulse of blood in your arteries. If it's this hard for binoculars, just imagine using a telescope with a much smaller field of view and lots more magnification. Humans don't make good telescope mountings. A good mounting should make pointing and steadying *easy and accurate*.

What makes a mounting good? A good mounting's capabilities lie somewhere between rock-solid steadiness and unrestrained looseness. If you want rock solid steadiness just cement your telescope to a boulder. Rock steady, but your telescope points in one and only one direction. If you want unrestrained pointing, just hold your telescope in your hands as already described. You can move it anywhere you want — it will dance around how it wants, but you never steady on anything. Finding the happy medium between pointing and steadying starts with restricting your freedom to point.

The Alt-Azimuth Mount

You only need two different types of motion to point anywhere. After all, that's what the human coordinate system is all about — although you have complete freedom to move any way you care, you restrict yourself (in thought, language and motion) to just up/down and left/right. So why not build a mounting that restricts your telescope the same way? Then name it an *alt-azimuth* mounting after altitude and azimuth, the two

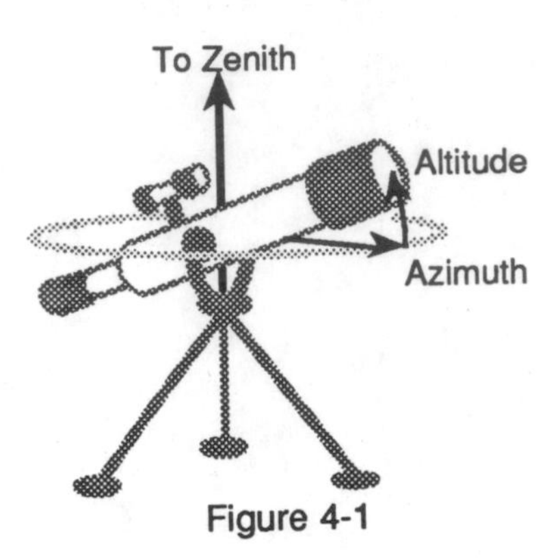

Figure 4-1

motions you've left it. The archetypical alt-azimuth mounting looks like Figure 4-1.

The feet of the tripod fix the mounting to the Earth, the rock of ages. The *zenith axis*, pointed at the zenith, connects the yoke to the tripod and lets the telescope pivot left or right around the zenith in azimuth. The *altitude axis*, crosswise to the zenith axis, connects the telescope tube to the yoke to let the telescope pivot up or down from the horizon — both motions in rigid imitation of you. For binoculars, the handiest alt-azimuth mount is a camera tripod. For astronomical telescopes, the best alt-azimuth mounting is the Dobsonian (to be discussed later). Most alt-azimuth refractors use a simple yoke mount like Figure 4-1. Regardless of their individual peculiarities, all alt-azimuth mountings share a common fault — the human coordinate system. Stars, after all, belong to the celestial sphere not to the human sphere.

Remember how the diurnal rotation of the Earth affects the appearance celestial sphere (Chapter 2); stars appear to move (obvious) but their motions in altitude and azimuth change in speed and direction (not so obvious). You can point a telescope at a star with an alt-azimuth mount and center it perfectly into your eyepiece — for an instant. The star continues on its diurnal path and you continue to chase it. Now hold it just a minute. Stars don't move that fast do they? When you go outside and watch, you don't see them whizzing across the sky. Yes, they do when your field of view is only a quarter degree and you've magnified their motions a hundred times.

Not only can a star whiz through your eyepiece, its direction and whizziness depend upon it's location in the sky (Chapter 2). From a northern-hemisphere chauvinistic viewpoint, a star in the northern sky moves in counterclockwise circles around the pole - its speed and direction in altitude and azimuth constantly varying - so you chase it by moving your telescope in counterclockwise circles around the pole, your speed and direction constantly varying. A star in the eastern sky moves upwards to the right, so you chase it upwards to the right. A star in the western sky moves downwards to the right, so you chase it downwards to the right. Of course, you could automate your telescope to track a star regardless of where it is in the sky.

That automation would require variable speed motors on each axis, a controller that continually varies each motor's speed and direction, and a computer to run the whole affair. Yet, for centuries, astronomers have tracked stars across the sky with a single clock drive. How? By devising mountings fixed to the celestial sphere, not the human sphere.

Box 4-1: Perception and Attention

If you've ever scanned the heavens under a really dark sky, you've been overwhelmed by all the stars. Yet, when you concentrate on a particular constellation, its stars seem to fade.

When you scan the heavens your attention isn't focused on any particular spot of your retina, so you perceive a general impression of what your whole retina sees, not just one particular spot. That perception is dominated by what your low light level sensors (rods) see, so you see better. When you stare at a star ,you center it *and* your attention on your fovea, the least light-sensitive part of your retina. So the star looks faint.

You can, however, shift your attention away from your fovea — the two are not inseparable. This process is called *averted vision*. You look at a star with a more sensitive part of your retina and shift your attention over to it from your fovea. When you do, details start to pop up.

The more attention you pay as you observe, the more detail you'll see, for your trained eye builds up a picture from the best bits and pieces of the image. As you perceive snatches of detail, add them to your sketch — use paper memory to help you remember the detail. You can use your sketch the other way, to refresh your trained eye on details. Above all, take your time in observing. Look for details and you'll find them.

The Equatorial Mount

Well, how do you tie your mounting to the stars? Start by leaning on your alt-azimuth mounting until the zenith axis tilts over to point at the north celestial pole. Voila! It's now a *polar axis* and you have an *equatorial mounting* (Figure 4-2), the official name for celestial mountings. Your equatorial mounting lives in celestial coordinates and hence your telescope now lives in celestial coordinates.

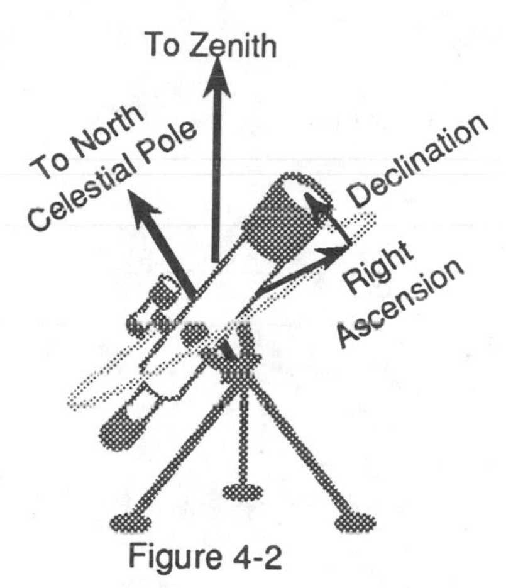

Figure 4-2

The polar axis lets your telescope revolve from east to west around the pole just like the stars do. The *declination axis*, the axis crosswise to the polar axis, lets your telescope swing north/south of the celestial equator. A dial on each axis lets you see where your telescope is pointed. The *right ascension dial* displays your telescope's current right ascension; the *declination dial* displays the current declination. Together they are called *setting circles* since you use them to *set* the telescope on a star. One last thing: the *latitude adjustment* lets you tilt the polar axis until it points at the north celestial pole. Why is it called a latitude adjustment? Because the altitude of the north celestial pole above your horizon depends upon your latitude (Chapter 2).

Since the stars revolve smoothly around the pole star once per *sidereal* (stellar) day, a clock motor attached to the polar axis will drive your telescope around the pole in perfect sync with the stars. The complexities you saw in following the motion of stars with an alt-azimuth mount disappear. However ,only your telescope has been transferred into celestial coordinates — you still live in human coordinates but now have to think, speak and act in celestial coordinates. It's time to get tilted.

Fortunately, your equatorial mounting helps you since it can only move in celestial coordinates — north/south and east/west. You can't move it wrong — up/down or left/right. Finding a star with an equatorial mounting is trivial, no more worrying about time of day or season of year to point at a star. Just look up its right ascension and declination and swing the telescope east/west till the right ascension dial displays the star's right ascension and north/south till the declination dial displays the star's declination. Bingo, you have the star in the eyepiece ready to observe. Make sure the clock motor is plugged in and the telescope will track the star all night long if need be. Nothing could be simpler, right? The catch? You have to install your equatorial mount into the celestial coordinate system before you can use it.

A working equatorial mounting's polar axis points at the north celestial pole. Where does your equatorial mounting's polar axis spend most of its time pointing? Probably towards your closet door or the floor of your car's trunk. Each time you take your telescope out of its resting place you must transport it from the earthly plane into the celestial sphere. Its polar axis is no celestial lodestone: it will not seek the north celestial pole of itself. Worse yet, there is no point marked "north celestial pole" on the celestial sphere for you to point its polar axis at. How do you point something you can't see through, at something you can't see? Some equatorial mountings now have alignment telescopes built right into the polar axis, but the majority don't.

Fortunately , for most purposes you can get by with a "quick and dirty" technique that puts your equatorial mounting within 1° of perfect alignment using the "*pole*" star *Polaris*. Look at the north polar star map and you'll notice an amazing coincidence. There's a second magnitude star about 1° from the north celestial pole. So if you point the polar axis at Polaris (which you *can* see), you've aligned your polar axis within 1° of the north celestial pole (which you *can't* see).

Quick and Dirty Polar Alignment

Step 0: *Set the Polar Axis Tilt* - getting ready.

NOTE: You perform this preparatory step once and seldom thereafter — whenever you drop your telescope or move to another latitude.

Make sure the finder aligns properly with the main telescope (Chapter 3) and set the latitude adjustment to your latitude. If you are at the north pole of the Earth (latitude 90°), the north celestial pole is straight overhead (90° altitude). If you live at the Earth's equator (latitude 0°), the north celestial pole lies on the horizon (0° altitude). If you live at latitude 30°, the north celestial pole lies 30° above the horizon. What can you infer from this? The altitude of the north celestial pole equals your latitude; you tilt your polar axis up from the horizon by the amount of your latitude.

Step 1: Point Telescope Parallel to the Polar Axis - Make the telescope a polar axis gun sight.

NOTE: How do you aim something you can't see through?

A) Move the telescope on its declination axis until the declination dial reads 90°. This points the telescope and its finder scope along the polar axis.

B) Lock the declination axis.

Step 2: Align Polar Axis to North Celestial Pole — Use the telescope as a gun sight.

A) Move the entire telescope, touching only the mounting by its base or tripod (don't touch the telescope tube) until the telescope points northward.

B) Watch through the finder and move the mounting by its base or tripod (don't touch the telescope tube) until you can see Polaris in the finder.

C) Continue to watch through the finder and move the mounting by its base or tripod (don't touch the telescope tube) left or right to position Polaris directly over or under the crosshair.

D) Raise or lower the polar axis tilt — adjust the latitude adjustment — until Polaris is on the crosshair.

NOTE: For most finders, the image is simply inverted, so if Polaris is above the crosshair you need to lower the polar axis tilt and if Polaris is below the crosshair you need to raise the polar axis tilt. If your finder has a diagonal, then these directions may be the same or reversed depending upon how the diagonal is oriented in the finder.

E) Repeat steps C and D as needed.

NOTE: You are now aligned on Polaris which is a degree off of the north celestial pole.

Step 3: Set Right Ascension Dial - Hitch your clock to a star.

Stars move and their right ascensions move right along with them. The clock motor that drives your telescope to track the stars also (for most but not all equatorial mounting designs) drives the right ascension dial to keep pace with the stars. What happens when you turn off the clock? The stars don't stop moving, but your right ascension dial does. Because it stops, you must reconnect it to the stars whenever you turn on the clock motor.

A) Turn the clock motor on.

B) Unlock the declination axis, and move the telescope tube (not the mounting) until it points at a bright star near the celestial equator (Procyon or Altair for example).

C) Center the bright star in the finder, then in the main telescope and lock the declination axis.

D) Rotate the right ascension dial (not the telescope) until it displays the correct star atlas right ascension for the bright star.

NOTE: The clock motors on some equatorial mountings (depending upon their design) only drive the telescope, not the right ascension dial — if you wait a few minutes, you may notice that the right ascension you've just set for the bright star is wrong by a few minutes! For this type of equatorial mounting, you supply the clock motor for the right ascension dial, keeping it adjusted manually, while the real clock motor drives the telescope to track the star. The easiest way to perform this duty (and a good practice in general) is always to check the right ascension dial just before leaving a star and make sure it displays the correct right ascension. If it doesn't, then set it to the correct right ascension.

Both advanced alignment techniques, which put the polar axis closer and then right on the north celestial pole, begin with the quick and dirty approach and so should you.

Box 4-2: Purchasing a Telescope

Lousy optics, a lousy finder telescope and a lousy mounting make a lousy deal for you; pretty packaging and pretty advertising make a pretty profit for someone else. There are a lot of paperweights out there masquerading as astronomical telescopes. So what do you trust? For starters, there are no astronomical telescopes for under $300, only paperweights (even if it has a smaller paperweight bolted onto the side and perches on the most elegant paperweight stand on the market). Your best source of advice is one of the many amateur astronomer groups listed in Appendix B. Contact one of these groups, go to one of their meetings, try viewing through different telescopes, and ask for help. You won't regret it.

What can you get for under $300? An excellent pair of binoculars. Even the most inexpensive pair of binoculars has far more capability than Galileo's telescope. Typical binoculars suitable for astronomical observing are 7x50's or 10x50's The first number states the magnification, the second number the objective lens diameter (aperture) in millimeters.

Remember that binoculars are refractors and subject to all the same aberrations. Powerful binoculars often have powerful aberrations. Still binoculars let you see the craters on the moon, the moons of Jupiter, star clusters and nebulae along the Milky Way and thousands of stars invisible to the unaided eye. When you're ready to invest further into the universe, then you're ready for an astronomical telescope.

One type of astronomical telescope you may consider is called a Dobsonian, after the amateur astronomer who popularized a sturdy, steady, easy to move and easy to build alt-azimuth mount. Yes, that's right, an alt-azimuth mounting not an equatorial mounting. Good equatorial mountings are as expensive as the telescope optics. Dobsonians are generally used at low powers so that the celestial objects don't whiz through the eyepiece. You have to track manually since they cannot be easily clock driven. Furthermore, because a Dobsonian lives in the human coordinate system, you cannot use celestial coordinates to find objects. You find objects just as you would with the naked eye — directions and distances to them from bright stars on the star maps tell you directions and distances from them in the sky.

You need an equatorial mounting if you're just learning or you need the assistance of right ascension and declination to find faint stars or you're performing astrophotography and need to track stars smoothly and evenly. What if you can find any object with one coordinate tied behind your back? What if you're not doing astrophotography? What if all you want to do is see faint objects? Then consider a Dobsonian and put the money you save on the mounting into bigger optics.

Coordinated Polar Alignment

The quick and dirty technique aligns your telescope to about 1° of the true north celestial pole, just within your main telescope's field of view (Project 4-1). True, you should still see the star you're after in the finder (thanks to its wider field of view) and can use the finder to center it into the main telescope. However, what happens when other errors, errors you don't even know you've run into yet, keep you from getting anything you want into the finder? Given the unknown and known sources of error, you should eliminate the known errors now and you will have less to worry about later. A good policy, for as you learn to recognize what leads to errors you also learn to eliminate them. Right now you know to eliminate that residual 1° error in polar alignment left over from the quick and dirty technique. How? Using the coordinated polar alignment.

Step 0: *Quick and Dirty Polar Alignment* - getting ready.

Start from where the quick and dirty ritual stops.

Step 1: Find Polaris by Its Coordinates — Point to Polaris' right ascension and declination .

A) Move the telescope in declination until the declination dial reads Polaris' declination. Lock the declination axis.

B) Move the telescope in right ascension until the right ascension dial displays Polaris' right ascension.

Step 2: Align Mounting to Center Polaris in Telescope — Align the polar axis to the north celestial pole.

A) Watch through the finder and move the mounting by its base or tripod (don't touch the telescope tube) left or right to position Polaris directly over or under the crosshair.

B) If Polaris is above the crosshair , you need to lower the polar axis tilt, if Polaris is below the crosshair, you need to raise the polar axis tilt (in other words, adjust the latitude adjustment).

C) Repeat steps A and B as needed.

Step 3: Check Right Ascension Dial - Check your starting point.

A) Unlock the declination axis and move the telescope tube (not the mounting) until it points at a bright star near the celestial equator (Procyon or Altair, for example).

B) Center the bright star in the finder, then in the main telescope and lock the declination axis.

C) Rotate the right ascension dial (not the telescope) until it displays the correct right ascension from The Trained Eye Star Atlas for the bright star.

Step 4: Repeat until Perfect — Repeat steps 1 through 3 until there is no change.

> *NOTE: You have now aligned the polar axis to as accurately as you can read your right ascension and declination dials.*

Perfect Polar Alignment

This last technique takes a lot more time but is well worth it for one purpose: astrophotography. Astrophotographs require long exposures, up to hours, so small tracking errors cause big problems. The perfect polar alignment technique works by contrasting the motions of the stars to the motion of your telescope. The more time you take the closer you come to the nirvana of perfect alignment. Unfortunately, the more time you take to align the less time you have to observe. You'll also need a high-power (short focal length) eyepiece that has a crosshair, otherwise known as a *guiding eyepiece*.

Step 0: Coordinated Polar Alignment Ritual - Get as close as you can first.

Step 1: Left/Right Adjustment - Point the polar axis directly north.

A) Find a bright star near the celestial equator and high in the sky (near your meridian) and get it centered in your main telescope with the guiding eyepiece.

B) Turn off your clock drive for a second and align one of the crosshairs with the direction the star drifts. That direction is west, the opposite is east, north is counterclockwise from west and south is counterclockwise from east. Remember these directions in your eyepiece. Plug your clock drive back in and recenter the star on the crosshair and watch.

C) If the star drifts south, shift the polar axis to the left. If the star drifts to the north, shift the polar axis to the right. It won't take very much shift. Keep watching and shifting left/right until the star doesn't drift north/south no matter how long you watch.

> *NOTE: Don't forget to reset the right ascension dial for the amount of time the clock motor was turned off.*

Step 2: Up/Down Adjustment — nirvana.

A) Now point your telescope at a star near the equator near the eastern or western horizon (about 20° altitude). Center it in your guiding eyepiece and watch.

B) If the star drifts south, raise the polar axis. If the star drifts north, lower the polar axis. You won't have to raise or lower the polar axis very much. Keep watching and raise/lower the polar axis until the star doesn't drift north/south no matter how long you watch.

C) Repeat steps 1 and 2 until you reach nirvana (perfect bliss or boredom).

> *NOTE: Don't forget to check your right ascension dial when you're done.*

Who's in Charge?

You meticulously align your equatorial mounting, carefully monitor the right ascension and declination dials as you *slew* the telescope (move the telescope from one point in the sky to another around the polar and/or declination axes) to the exact coordinates of your star, rush to the eyepiece and see ... nothing. Why didn't it work? Because you trust the telescope, the machine, more than yourself. A telescope will not forgive your mistakes anymore than an automobile will.

Think back to driver's education. Nothing was intuitive, nothing was second nature. Why didn't you just trust the car to do the right thing? You controlled the machine with conscious effort, thinking through each action. Now you drive home from school or work, pull into the garage, turn off the ignition and don't remember anything about the drive. You follow the usual route, make the usual lane changes, stop at the usual traffic lights and turn at the usual corners — all intuitively. Using your equatorial mounting will become just as intuitive — but not yet. Right now you have to think through each action, learn to control the machine by learning to check its veracity.

You don't believe everything you read ,so why do you believe everything the declination and right ascension dials display? There are many ways they can be wrong; you might have aligned on some other star than Polaris, your declination dial may be out of whack (your telescope doesn't point along the polar axis when the declination dial displays 90° declination), you may have used the wrong right ascension for the bright star when setting the right ascension dial, someone may have accidently spun your right ascension dial, you may have forgotten to update your right ascension dial before leaving the last star. So don't just stand there in consternation — check yourself!

Step back and look where the telescope is pointed. If you're trying to find a star in Orion, then the telescope should point towards Orion. If Orion is in the west and your telescope points east, you know you have a problem. If it looks like its pointed in the right general direction, you have to look a little closer.

Go back to a bright star you can find easily. Make sure that when you have it on the finder's crosshairs that it also centers in the main telescope. If not, tweak the finder's alignment until its crosshairs faithfully indicate where the main telescope points. Now you're ready to check the equatorial mounting.

Compare the star's coordinates to what the right ascension and declination dials display. If the right ascension is wrong, correct it by dialing up the correct right ascension. If the declination dial is way off (more than 3°), then go back to square zero and redo the polar axis alignment (Chapter 4). You might even want to go back to square minus one and check the declination dial's adjustment.

If the declination dial is slightly misaligned (less than 3°), then just keep track of the *declination error*. If the declination dial reads a degree high, you know to aim a degree high. If it reads a degree low, you know to aim a degree low. For those who like equations (or at least don't fear them):

Declination Error = Telescope declination - Star Atlas Declination

Declination error does have one peculiarity: different values in different parts of the sky. This really isn't news since you use this to advantage in the perfect polar alignment technique, but it does affect finding stars. You can't use the declination error from one half of the sky to find a star in the other half of the sky.

Even with perfect polar alignment, you should still question the right ascension and declination dials by rechecking them on each celestial object you find. Each time you check and the dials jibe, you reinforce your self-confidence; you reassert your control over the machine.

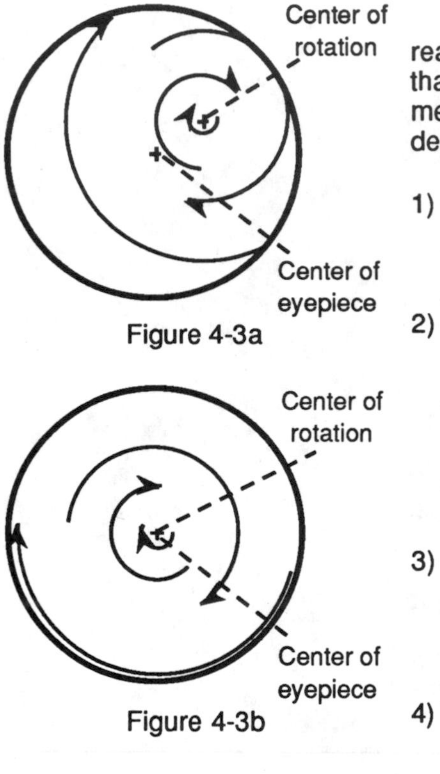

Figure 4-3a

Figure 4-3b

If the declination dial is still grossly misaligned (more than 3°) after realigning the polar axis, then adjust the declination circle. Keep in mind that this has nothing to do with polar alignment. You're just making a mechanical adjustment to your equatorial mounting so that when the declination circle displays 90° it indeed points along the polar axis.

1) Move your telescope to 90° declination and shift the mounting until the telescope points at some distant terrestrial object (tree, building, mountain, etc.) .

2) With the object centered in the finder, whip the telescope around in right ascension from one side to the other while watching through the eyepiece. If the telescope is truly aligned with the polar axis, then the object should stay centered and you will see it rotating about the center of the eyepiece. If it is not perfectly aligned, then it will rotate about some other rotation point in the eyepiece that really is where the polar axis points. (Figure 4-3a).

3) Move the telescope in declination until the rotation point is at the center of the finder and again whip the telescope around in right ascension. The object should circle about the center of the eyepiece. (Figure 4-3b)

4) You've now found where 90° declination should be. So loosen the set screw on the declination dial and rotate the dial on the declination axis (make sure the axis doesn't turn) until the dial displays exactly 90°. Cinch down the set screw and you're done.

5) Now you're ready to redo the polar alignment technique of your choice.

All the skills described so far, whether learning how to align your equatorial mounting or how to confirm its proper working, are simple. That may not seem so right now because you are confusing simple with easy. As yet you haven't learned these skills. They aren't intuitive — they aren't easy. That comes only with practice.

Project 4-1
Field of View and Light Gathering Power

Before you begin, set up and align your telescope using the quick and dirty technique. The three main steps are:

- Step 1: Point Telescope Parallel to the Polar Axis
- Step 2: Align Polar Axis to North Celestial Pole
- Step 3: Set Right Ascension Dial

Part A: Field of View

You now get to measure the angular sizes of the fields of view for your finder scope, low-power eyepiece and high-power eyepiece by measuring the time it takes for a star to cross each field of view and converting it back to degrees.

1) Find a reasonably bright star near the celestial equator (declination near 0°). One several hours away from your meridian will be easier to observe. Get the star centered in the finder scope. Turn off the clock drive and measure the time for the star to drift from the center to the edge of the field. Double this value to get the edge to edge drift time and record it in Table 1.

2) With the same star and the same procedure, measure the time required for the star to move all the way across the field of the low-power eyepiece. Be sure the star drifts right through the center of the eyepiece. Do this two times, take the average and record your result (don't double it like you did for the finder) in Table 1.

3) Repeat the above for the medium- and high-power eyepieces.

4) Since the Earth rotates 360 ° in 24 hours, or 1° every 4 of time, a star on the celestial equator will drift 1° in the sky every four minutes of time. Use this relationship between drift time and degrees to convert the time it takes the star to drift across your eyepiece to the field of view of your eyepiece;

    4 minutes (m) of TIME = 1 degree of ARC (°)
    4 seconds (s) of TIME = 1 minute of ARC (')
    1 second (s) of TIME = 15 seconds of ARC (")

Now convert the drift times to fields of view in °, ', or " (not time) and record the results in Table 1. Examine your results (Table 1), answer questions 1, 2 and 3 and summarize what you've learned in your writeup.

Table 1

Optical Device	Drift Time min sec	Field of View ° ' "
Finder Scope		
Low Power (≈40mm)		
Med Power (≈20mm)		
High Power (≈10mm)		

Questions

1) What does your measured "field of view" really mean in terms of how much of the sky you can see through the eyepiece?

   Hint: How many degrees is a hand? How many degrees is a finger?

2) What is the advantage of using a low-power eyepiece for finding a star?

3) Why use a high-power eyepiece at all?

Part B: Limiting Visual Magnitude

Get comfortable, flashlights off, take that glint out of your eyes and get them dark adapted! For your unaided eye observations, you will identify stars from The Trained Eye Star Atlas and find the faintest star you can see in a constellation. Once you can identify the brighter stars in a constellation, you can use the patterns of stars on The Trained Eye Star Atlas to identify fainter stars and know where to look for them. If you see it, then try again with an even fainter star. You will perform this experiment for a constellation high overhead and one low on the horizon.

For your telescopic assisted eyes, you will work with stars in a cluster in much the same way. You will point the telescope towards either the Pleiades or Praesepe star cluster (whichever is up). The Pleiades is in Taurus and the Praesepe is in Cancer Look up their right ascension and declination in The Trained Eye Star Atlas. You'll find finding charts for both clusters in Project 4-2. You will use the cluster you selected to determine the limiting magnitudes of your finder scope and the telescope for several different eyepieces.

1) Unaided Eye

A) Pick a constellation overhead that has lots of bright and faint stars.

B) Identify a 2.0 to 3.0 magnitude star from The Trained Eye Star Atlas and look for it.

C) If you find the star, then repeat the process for one about 0.5 magnitudes fainter.

D) Keep going fainter by about 0.5-magnitude increments until you identify the absolutely positively faintest star you can just catch glimmering. Write down its Bayer-Flamsteed name and apparent magnitude in Table 2. This is your limiting magnitude for the unaided eye.

E) Measure its altitude and azimuth in the usual manner.

F) Repeat steps A through E for a constellation in the darkest patch of sky near the horizon (good luck if you live in a big city).

Table 2

Star Name	Limiting Magnitude	Altitude °	Azimuth °

Use your observations to answer questions 7, 8 and 9.

2) Finder Telescope

You may find it hard to identify individual cluster stars because of the small scale (wide field of view and low magnification). You can estimate the limiting magnitude by just counting how many stars you see in the cluster. If you count five stars, then you are seeing just the five brightest stars in the cluster, so your limiting magnitude is approximately that of the 5th brightest star in the cluster. That's the ideal case, however. Table 2 in Project 4-2 list just the standards in the cluster; so you have to make allowance for the unknowns. Do that by subtracting one star for every two you count. So if you count 10 stars, then you are seeing the 10 brightest stars, which include five standards and five unknowns. So subtract the five unknowns and look for the 5th standard down in Table 2 of Project 4-2. Record that limiting magnitude in Table 3.

3) Main Telescope

Using the main telescope, you can identify individual stars. Start with the low-power eyepiece. Locate the faintest star you can see in the cluster and identify it on the cluster's finding chart (that's the picture of the cluster with little numbers written next to each star). The little number written next to the star is its star id number. Keep track of it. Now find a star that's even fainter. Remember, you're trying to find the absolutely positively faintest star you can see in the cluster. Identify it from the finding chart, look up its magnitude from the Table 1 of Project 4-2, and record your results in Table 3 of this project. Repeat this process for the medium-, and high-power eyepieces.

If you can see all the standards on the list, then the list just doesn't go faint enough for you to establish your limiting magnitude. You should still look for the effects of changing magnification. Reexamine which eyepiece made seeing the faintest star easiest.

Hint: Notice how dark the sky looks through each eyepiece and how well the stars stand out.

Table 3

Optical Device	Star Name	Limiting Magnitude
Finder Scope		
Low Power (≈40mm)		
Med Power (≈20mm)		
High Power (≈10mm)		

Use your observations to answer questions 10 through 15. Discuss your results and their significance in your writeup.

Questions

4) How many magnitudes fainter can you see overhead than near the horizon?

5) How many times fainter are you seeing overhead?

Hint: Use the table in Box 1-2 to convert the magnitude difference between the two stars in Table 2 to a brightness ratio.

6) Why can you see fainter stars overhead than near the horizon? List and discuss the probable causes.

7) How many magnitudes fainter do you see with the finder telescope than the unaided eye? How many times fainter are you seeing?

Hint: Use the table in Box 1-2 to convert the magnitude difference between the two stars into a brightness ratio.

8) How many magnitudes fainter do you see with the main telescope than with the unaided eye? How many times fainter are you seeing?

Hint: Use the table in Box 1-2 to convert the magnitude difference between the two stars into a brightness ratio.

9) How many magnitudes fainter do you see with the main telescope than with the finder telescope? How many times fainter are you seeing?

Hint: Use the table in Box 1-2 to convert the magnitude difference between the two stars into brightness ratio.

10) Contrast your limiting magnitudes for your eye, finder telescope and main telescope with the limiting magnitudes for various apertures listed in Table 3-1, Chapter 3. Explain why your results differ.

11) What happens to the brightness of the sky as you switch to higher and higher power eyepieces? Describe.

12) For your main telescope, did you see fainter with higher power or lower power eyepieces? Explain.

Project 4-2
Visual Photometry Of Stars

In this project, you will estimate the visual magnitudes of stars as seen through a telescope. Attached are finding charts for the brighter stars in the open star clusters named Pleiades and Praesepe. The Pleiades is in the constellation Taurus, the Praesepe is in the constellation of Cancer. Their right ascensions and declinations can be found in The Trained Eye Star Atlas. You can work with one of these two clusters. Each star in a cluster is identified by a two to four digit "ID" number generally written below the star it identifies (but not always). Some star numbers begin with the letter F, indicating that these are "field" stars, which means they are not members of the cluster, they're just in the way.

The procedure for determining visual magnitudes through your telescope is identical to the technique you used in Project 1-1 to determine visual magnitudes from the slide. Now you get to work with real stars. For your convenience (in case you've misplaced project 1-1), here is a repeat of how to perform visual photometry.

Visual photometry is the process of determining the brightnesses of unknown stars by comparing their brightnesses to nearby standard stars of known brightness, this comparison being made with your eye. The standard stars that match the unknown the closest in brightness set upper and lower limits on what the visual magnitude of the unknown star can be. That is, if the unknown is fainter than unknown A, but is brighter than standard B, then its visual magnitude is somewhere between those of A and B.

Attached are sheets with standard stars and unknown stars for each cluster. The standard stars are arranged in two ways: In Table 1, they are listed by ID number; in Table 2, by V magnitude. Use Table 1 to find the V magnitude of a particular star, use Table 2 to find a star with a particular V magnitude.

Follow the process described below for the unknowns in the cluster you have been assigned. When you have finished all the unknowns, record your results in Tables 3. Then take your data from Table 3 and list them from brightest to faintest (small mag to large mag) in Table 4. That will help you review your results to see if they are internally consistent. That is, if you say unknown U is brighter than standard A, but unknown V is fainter than standard A, then your results should show unknown U brighter than unknown V. If not, you have problems.

Example

Suppose you want to determine the visual magnitude of star U (a fictional unknown star). First find U on your finding chart, then locate it in the cluster. Next identify a standard that is near U in the cluster and on your finding chart — for example A. Compare the two. Which is brighter? Keep track of your observations by making notes like the following:

Unknown U is fainter than standard A of V mag 2.87.

Since you know U is fainter than 2.87 magnitude, now compare it to a standard that is fainter than 2.87 magnitude, like B, and record your results:

Unknown U is brighter than standard B of V mag 8.69.

You have now narrowed U's magnitude down to somewhere between 2.87 and 8.69. Continue to work on narrowing it down. For example, compare it to standard C.

Unknown U is brighter than standard C of V mag 4.31.

You have now narrowed U's magnitude down to somewhere between 2.87 and 4.31. Now compare to standard D.

Unknown U is slightly brighter than standard D of V mag 3.88.

So you now know U is somewhere between 2.87 and 3.88 in magnitude, but probably pretty close to 3.88. Pretty good so far, but the range is still too large. Now, find another standard slightly brighter than D and compare U to it.

Unknown U is slightly fainter than standard E of V mag 3.64.

At this point, you should have narrowed the visual magnitude of U to the 3.64 to 3.88 magnitude range. Now recompare U to both of these stars. Which is it closer to in brightness? We could call it 3.76 if it appears halfway between, or 3.70 if it is closer to the 3.64 magnitude standard, or 3.80 if it is closer to the 3.88 magnitude standard. Such a thought train would look like the following:

Unknown U is slightly fainter than standard E of V mag 3.64.
Unknown U is slightly brighter than standard D of V mag 3.88.
Unknown U is closer to standard E.
I guess Unknown U has a V magnitude of 3.70.

Tricks of the Trade

1) Use Table 1 when you want to find the V magnitude of a particular star.

2) Use Table 2 when you're trying to find a star of a particular magnitude.

3) Don't zero in on your star too fast by visually picking one that at first glance seems to match - you won't be that lucky at the telescope.

4) The last two standards that bracket each unknown should not differ by more than 0.5 magnitudes.

5) Look for small patterns of stars (triangles, squares, pentagons, etc.) to find your way around the slide.

6) Try using higher power eyepieces on the fainter stars.

Use your results to answer the following questions. Be sure to include your comments, impressions, problems, ideas in your writeup.

Questions

1) Approximately how many times brighter is the brightest unknown star than the faintest?

Hint: Use Box 1-2 (Chapter 1) to convert magnitude difference to brightness ratio.

2) What property of the star images are you using to estimate their brightness?

3) List and discuss the sources of uncertainty (things that give you trouble) in performing visual photometry.

Pleiades

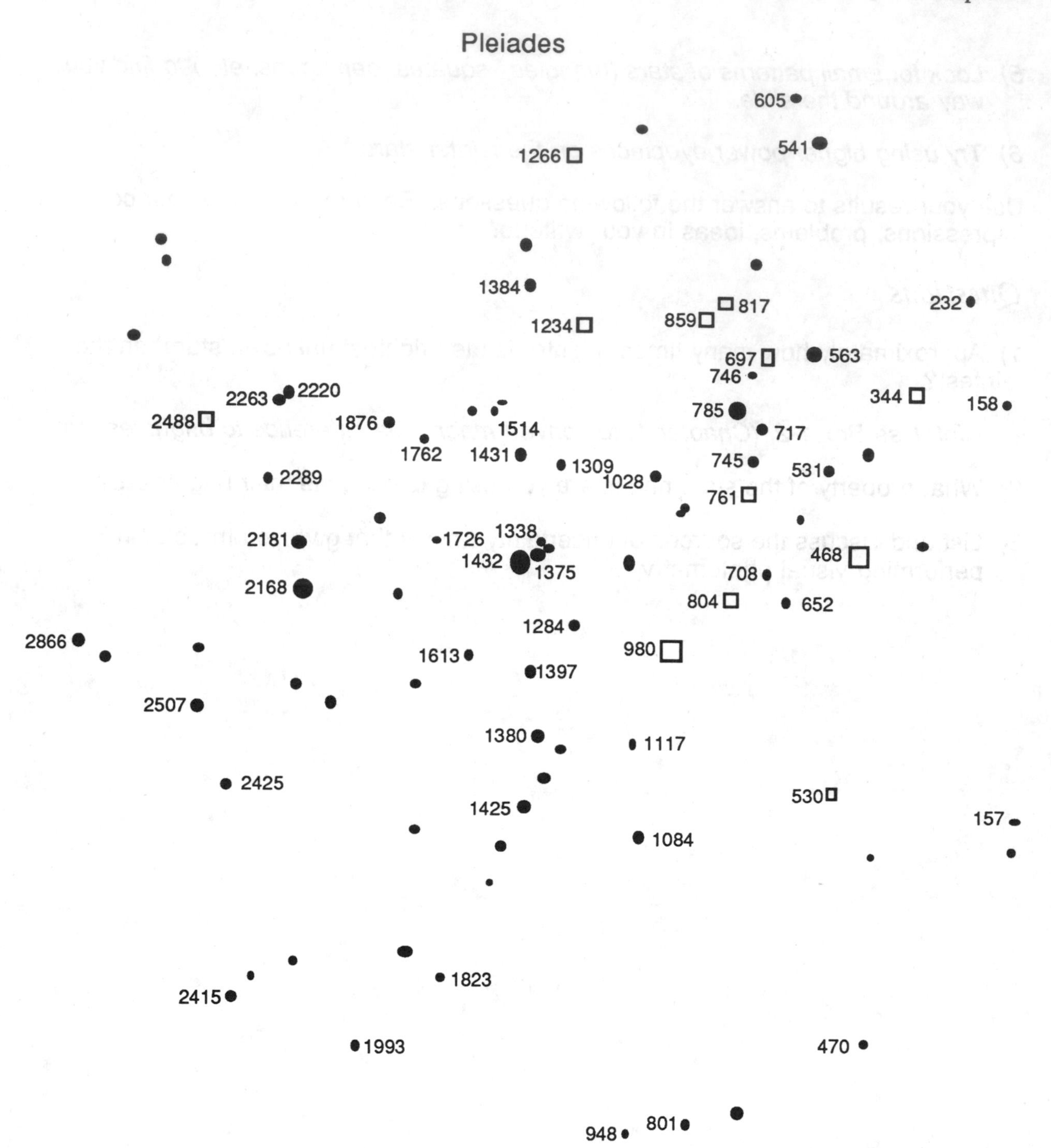
605
541
1266
1384
232
817
1234
859
697
563
746
2220
2263
344
158
2488
1876
785
1514
717
1762
1431
1309
745
531
2289
1028
761
2181
1726
1338
1432
1375
468
708
2168
804
652
1284
2866
980
1613
1397
2507
1380
1117
2425
530
157
1425
1084
1823
2415
1993
470
948
801

Pleiades

Table 1: Standards ordered by star

Star #	Visual Magnitude
157	7.90
158	8.23
232	8.06
470	8.95
531	8.58
541	5.64
563	4.31
652	8.04
708	10.13
717	7.18
745	9.45
746	11.27
785	3.88
801	6.85
1028	7.35
1117	10.20
1084	8.11
1284	8.37
1309	9.46
1338	8.69
1375	6.29
1380	6.99
1384	7.66
1397	7.26
1425	7.77
1431	6.81
1432	2.87
1514	10.48
1613	9.88
1726	9.25
1762	8.27
1823	5.45
1876	6.95
1993	8.37
2168	3.64
2181	5.09
2220	7.52
2263	6.60
2289	7.97
2415	8.10
2425	6.17
2507	6.74
2866	6.93

Table 2: Standards ordered by V mag

Star #	Visual Magnitude
1432	2.87
2168	3.64
785	3.88
563	4.31
2181	5.09
1823	5.45
541	5.64
2425	6.17
1375	6.29
2263	6.60
2507	6.74
1431	6.81
2866	6.93
801	6.85
1876	6.95
1380	6.99
717	7.18
1397	7.26
1028	7.35
2220	7.52
1384	7.66
1425	7.77
157	7.90
2289	7.97
652	8.04
232	8.06
2415	8.10
1084	8.11
158	8.23
1762	8.27
1284	8.37
1993	8.37
531	8.58
1338	8.69
470	8.95
1726	9.25
745	9.45
1309	9.46
1613	9.88
708	10.13
1117	10.20
1514	10.48
746	11.27

Pleiades

Table 3: Unknowns ordered by star #

Star #	Visual Magnitude
344	
468	
530	
697	
761	
804	
817	
859	
980	
1234	
1266	
2488	

Table 4: Unknowns ordered by V mag

Star #	Visual Magnitude

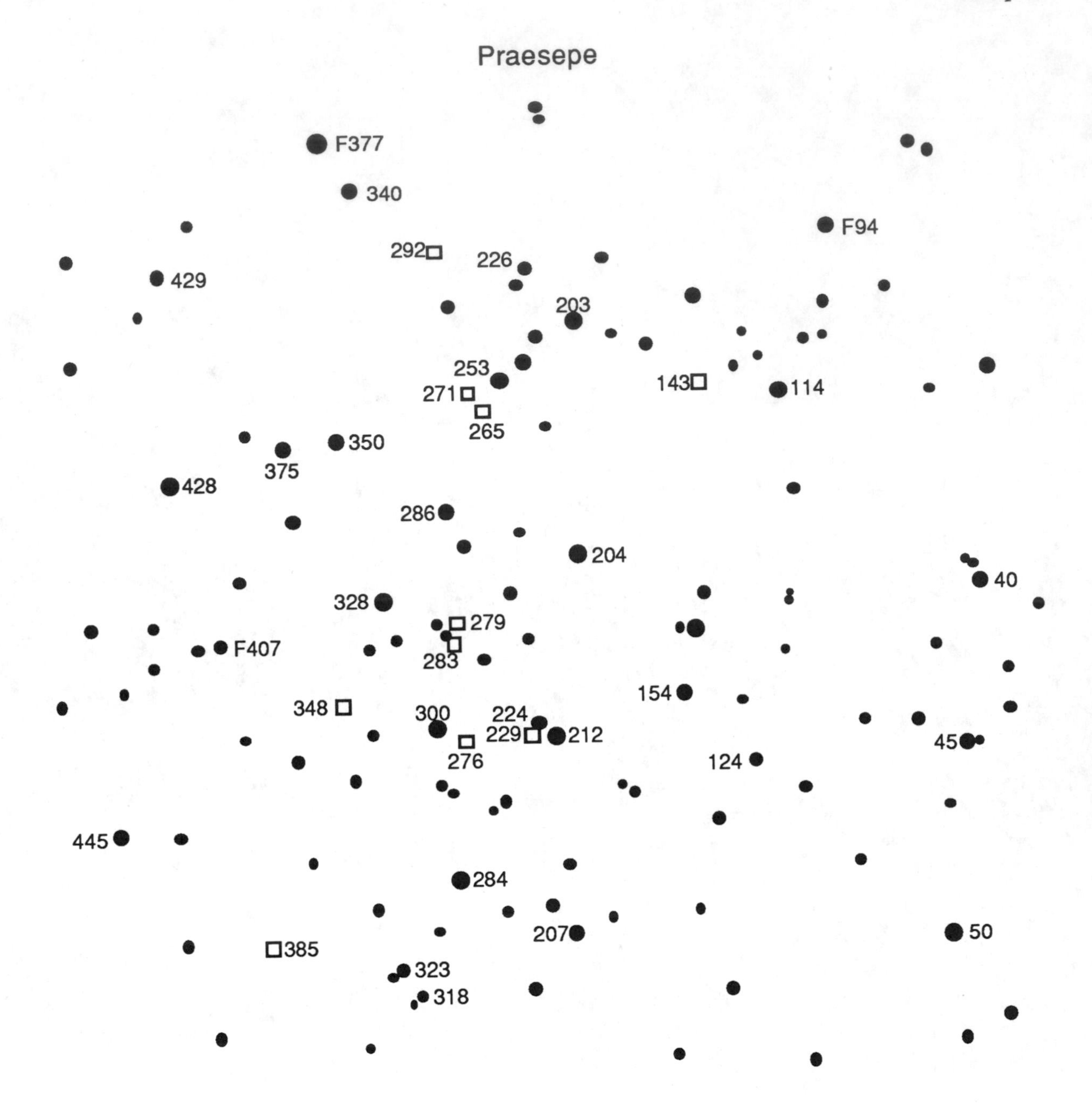
Praesepe
F377
340
F94
292
226
429
203
253
143
114
271
265
350
375
428
286
204
40
328
279
F407
283
154
348
300
224
229
212
45
276
124
445
284
50
207
385
323
318

Praesepe

Table 1: Standards Ordered by star #

Star #	Visual Magnitude
45	8.25
50	6.75
114	8.14
124	9.00
154	8.50
203	7.73
204	6.67
207	7.67
212	6.59
224	7.32
226	8.89
253	6.39
284	6.78
318	8.65
323	7.80
328	6.85
340	8.48
350	8.71
375	8.33
428	6.90
429	8.53
445	7.96
F94	7.83
F377	6.73
F407	8.50

Table 2: Standards Ordered by V Mag

Star #	Visual Magnitude
253	6.39
212	6.59
204	6.67
F377	6.73
50	6.75
284	6.78
328	6.85
428	6.90
224	7.32
207	7.67
203	7.73
323	7.80
F94	7.83
445	7.96
114	8.14
45	8.25
375	8.33
340	8.48
154	8.50
F407	8.50
429	8.53
318	8.65
350	8.71
226	8.89
124	9.00

Prasesepe

Table 3: Unknowns Ordered by star #

Star #	Visual Magnitude
143	
229	
265	
271	
276	
279	
283	
292	
348	
385	

Table 4: Unknowns Ordered by V mag

Star #	Visual Magnitude

Chapter 5
Taking the Measure of a Star

Look up at the stars, and the same message greets you that greeted Hipparchus two thousand years ago; twinkling points of light scatter across the hemisphere of night, each discernable from all the others by visual magnitude and sometimes a hint of color. Use a telescope and you'll see more stars, but no additional detail, just magnitude and color. What are stars? Perhaps they are objects like our Sun diminished to faint pinpoints by unimaginably great distances as the astronomer Aristarchus proposed over 2000 years ago. Equally ancient is the supposition that stars are holes in the celestial sphere, letting pale flickers of glory shine through, providing mere stage dressing for the god/planets in their passion play. Both suppositions are equally valid unless you know how to read the message in starlight. A message encoded in magnitude, position and color that carries the measure of stars.

How you read that message depends upon your conception of the universe — science and philosophy are inexorably linked. If you suppose the heavens to be magic and the Earth mundane, then stars are pixie dust and magnitude and color are meaningless. If you believe that the heavens and Earth share a common physics, then you have the philosophical Rosetta stone with which to read the message in starlight, that *stars are suns.* It took almost two centuries for this philosophical conception, a product of the philosophical and scientific revolution that spanned from Copernicus to Newton, to overturn the magical/mundane dogma.

It was a revolutionary concept, one that people died for, sprouting in the cracks that developed between religious dogmas during the Protestant reformation. Copernicus spun the philosophical world into motion by conceiving a *heliocentric* (Sun centered) universe and setting the Earth, here-to-fore fixed and immovable at the center of the universe, circling about the Sun. Recognizing the theological impact of his conception, Copernicus delayed its publication until after his death (by natural rather than theological causes). Martin Luther roundly condemned Copernicus' heresy (in rare agreement with the Catholic church). In 1600 the Catholic Inquisition burned Giordano Bruno at the stake for espousing, among other things, that stars are suns and as a gentle reminder to all philosophers since Bruno's conception was shared by others. If Copernicus could dethrone the Earth from the center of the creation, then why should the Sun be granted that favor in its place?

Tycho Brahe, the greatest observer of his time, compiled exacting observations of the planets for several decades in anticipation of proving Copernicus the fool. In 1609 Johann Kepler, the right mathematician in the right place at the right time, absconded with Tycho's legacy and, through insight, reason, analysis, and luck,discovered in Tycho's observations the true laws of the earth's and other planet's motion about the Sun. Galileo, a contemporary and confidant of Kepler, transformed the telescope from a toy into a scientific instrument and used his telescopic discoveries to proselytize tirelessly in support of Copernicus' theory, Kepler's ellipses, and Bruno's suns. Like Giordano Bruno, Galileo too aroused the interest of the Catholic Inquisition. Unlike Bruno, Galileo survived, but banned from celestial studies, he returned to terrestrial physics. Newton, born the year that Galileo died, united celestial and terrestrial physics into a common truth, culminating in 1687 the revolution started by Copernicus — the heavens and Earth do share a common divinity and a common physics. The stars are suns, the Sun a star, and all are mundane. Mundane because you can describe the stars, take their measure, in terms of the Sun, and measure both in terms of mundane (earthly) physics.

The Measure of a Sun

Describe the Sun and you've described the archetype for all stars. Our Sun is a hot glowing sphere of something. Hot implies *temperature*. Another word, *luminosity* measures the quantity of sunlight, how much energy per second the Sun radiates from its surface. *Luminosity* also measures a star's energy consumption — energy that has to be supplied somehow from something. Look at the sun's form and you find it a sphere. A sphere needs only one parameter to describe it — its *radius*. That takes care of "hot", "glowing" and "sphere", but what about "something"? The Earth is held in its orbit by the sun's gravitational field which implies that the something is matter. How much matter is the sun's *mass*, while *chemical composition* describes the kinds of matter. The quantity and kind of matter describe the fuel for a star's thermonuclear furnace — how it creates its *luminosity*. The list of descriptive parameters grows more detailed, but we end it here with the most fundamental: *temperature, luminosity, radius, mass*, and *chemical composition — all* concepts from terrestrial physics.

The ABCs of Light

Light transports more than just energy across great distances, light carries *information* transcribed in magnitude and color. Light spells out temperature, luminosity, radius, mass, and chemical composition for stars. So what is light? We beg the question by describing how light behaves — not what it is. In certain experiments light acts like a billiard ball, in other experiments like an ocean wave. So sometimes we call light a wave, sometimes we call it a particle. By assuming a wave nature for light you can explain color. But how do you describe a wave?

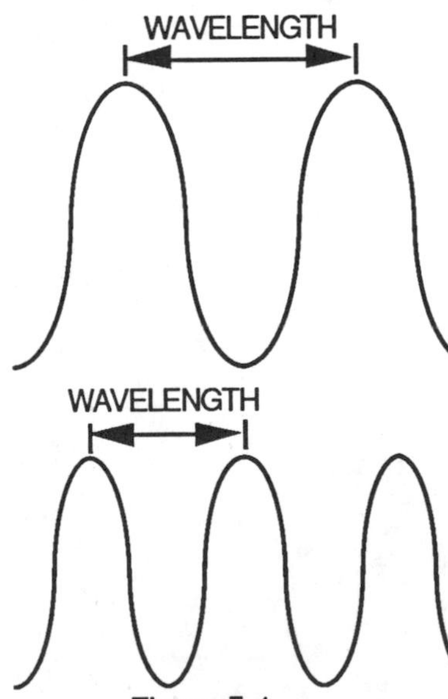

Figure 5-1

All waves, including light, can be described as a series of undulations (Figure 5-1) — a series of crests and troughs. The property of a wave that distinguishes one from another is *wavelength*; the distance from crest to crest (or trough to trough). The *wavelengths* of visible light are tiny, so small we measure them with a comparably small unit of measurement called an Ångstrom (abbreviated Å) which equals 1/10,000,000,000 of a meter or about 0.000000004 inches. Of course you don't percieve wavelength directly, what you see is different *colors of light* (In contrast to *color of an object*, the eye's estimate of which *color of light* is most intense coming from an object). What you percieve as a single *color of light* covers a wide range of wavelengths, a wide range of hues.

Violet	4000	to	4500	Å
Blue	4500	to	5000	Å
Green	5000	to	5500	Å
Yellow	5500	to	6000	Å
Orange	6000	to	6500	Å
Red	6500	to	7000	Å

There are also additional unseen colors — wavelengths you perceive with your skin, not your eyes. Hold your hand out to a heat lamp and you feel the infrared color of light. Lie on the beach too long and your sunburn is evidence of another invisible wavelength — the ultra-violet. The visible wavelengths compose only a small fraction of the true colors of the rainbow, the spectrum of *electromagnetic waves*:

Gamma Ray	0.1	Å	and shorter	
X rays	0.1	to	10	Å
Ultra-violet	10	to	4000	Å
Violet	**4000**	**to**	**4500**	**Å**
Blue	**4500**	**to**	**5000**	**Å**
Green	**5000**	**to**	**5500**	**Å**
Yellow	**5500**	**to**	**6000**	**Å**
Orange	**6000**	**to**	**6500**	**Å**
Red	**6500**	**to**	**7000**	**Å**
Infrared	7000	to	50,000	Å
Radio	50, 000	Å	and longer	

Electromagnetic waves, unlike ocean waves, are not oscillations of a physical medium — water sloshing up and down — they are self propagating oscillations of *electric* and *magnetic* fields. Another property that these different wavelengths have is different energies.

Light waves, unlike ocean waves, don't stretch out in a endless alternation of crests and troughs from source to destination. Light comes *quantized*, packaged into discrete wavelets, called *photons.* Each photon carries, at the speed of light, a discrete amount of energy set by its wavelength:

$$\text{Energy} = \frac{0.00000000000000048}{\text{wavelength}}$$

Put in wavelength (Ångstroms) and out comes energy in Calories. The importance of this equation is not the units of wavelength or energy, but the relationship between energy and wavelength:

shorter wavelength photons have higher energy
longer wavelength photons have lower energy

Violet photons have shorter wavelengths than yellow photons and therefore have higher energy. Red photons have longer wavelengths than both violet and yellow photons and correspondingly lower energies. It is this difference in energy between photons that the eye uses to discriminate their wavelengths, not the wavelengths themselves.

You have three types of color receptors on your retina each requiring a different energy range of photons to generate the electrical stimulus that your mind perceives as color: some of the color receptor cells require high energy violet-blue wavelength photons to trigger their chemistry, some are sensitive to medium energy green-yellow wavelength photons, and others require only low energy orange-red wavelength photons. Your brain compares the strength of the stimuli from each of these different color receptor cells to deduce the color of the object that emitted the photons. If that object glows from its own heat, then its color tells you its temperature.

The Temperature of Light

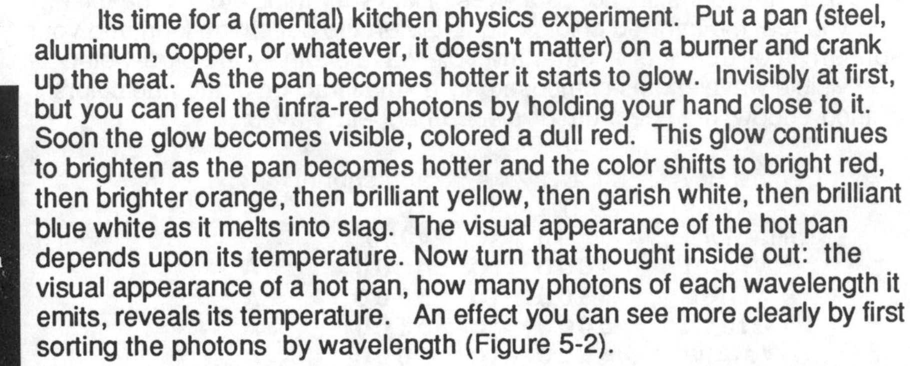

Its time for a (mental) kitchen physics experiment. Put a pan (steel, aluminum, copper, or whatever, it doesn't matter) on a burner and crank up the heat. As the pan becomes hotter it starts to glow. Invisibly at first, but you can feel the infra-red photons by holding your hand close to it. Soon the glow becomes visible, colored a dull red. This glow continues to brighten as the pan becomes hotter and the color shifts to bright red, then brighter orange, then brilliant yellow, then garish white, then brilliant blue white as it melts into slag. The visual appearance of the hot pan depends upon its temperature. Now turn that thought inside out: the visual appearance of a hot pan, how many photons of each wavelength it emits, reveals its temperature. An effect you can see more clearly by first sorting the photons by wavelength (Figure 5-2).

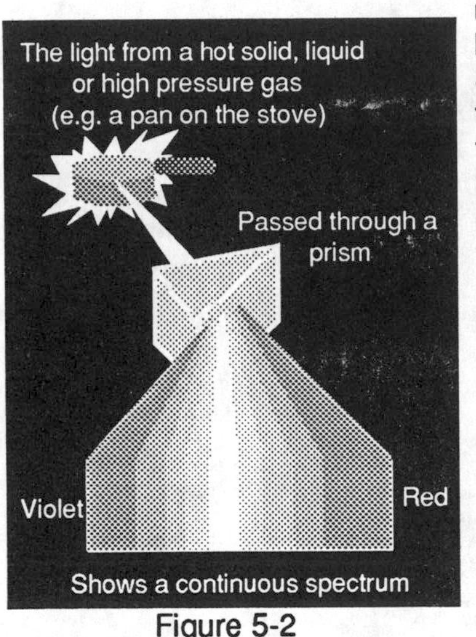

Figure 5-2

Pass the light from the hot pan through a prism and it will produce a continuous rainbow of colors (of wavelengths) from red to violet called a *continuous spectrum.* Regardless of the pan's composition, you'll see the same colors in the same order. Now crank up the heat and watch what happens to each and all wavelengths of light.

As the pan's temperature increases, you'll see each color of light grow more intense, you see more photons of all wavelengths. You'll also see that the color that glows the brightest, the wavelength at which the most photons are emitted, continually shifts to shorter wavelengths. To sum up your experiment, as the pan's temperature increases:

1) It glows brighter — it emits more photons of all wavelengths.
2) It becomes bluer — the wavelength at which the most photons are emitted shifts to shorter wavelengths.

per square inch.

This last caveat is of fundamental importance. The pan glow brighter, not because it gets bigger and so there's more of it to emit photons, but because each square inch emits more photons. Two simple rules, relating the color and brightness of a hot object, its light, to its temperature. Two simple rules that you can extrapolate to the stars — if you've ever seen a rainbow.

A rainbow is just the spectrum of the Sun cast by prisms of raindrops. Measure the intensity of all the colors in the rainbow and you can calculate the Sun's temperature (Rule 1). Just observe the color of the Sun, the color of any star , and rule 2 gives its temperature from blue hot to cool red .

The Magnitude of Luminosity

Regardless of its wavelength, the light from a star shines outward in all directions in a vain attempt to fill infinity. The mathematical statement of that relationship between the distance light travels from its source and how it spreads out is called the *inverse square law* :

$$\text{apparent brightness} = \frac{\text{luminosity}}{4 * \pi * \text{distance}^2}$$

It's called the *inverse square law* because the star's apparent brightness decreases as the distance squared just as common sense dictates. If the inverse square law was not encoded in common sense, then you couldn't cross the street at night, you wouldn't be able to tell a candle from a searchlight or headlights from stars.

In everyday life you don't measure a light's luminosity — count the number of photons it emits. After all, the eye doesn't excel at absolutes, it excels at comparisons. You compare its apparent brightness at one distance to its apparent brightness at some other distance: move a light twice as far away and it looks 4 times fainter, bring it 10 times closer and it looks 100 times brighter. This everyday version of the inverse square law looks like:

$$\text{brightness ratio} = \frac{1}{\text{distance ratio}^2}$$

and is shown in Table 5-1.

Table 5-1 Brightness Ratio and Distance

Brightness Ratio		Distance Ratio	
10,000	times brighter	100	times closer
100	times brighter	10	times closer
16	times brighter	4	times closer
9	times brighter	9	times closer
4	times brighter	4	times closer
1	unchanged	1	unchanged
4	times fainter	2	times farther
9	times fainter	3	times farther
16	times fainter	4	times farther
100	times fainter	10	times farther
10,000	times fainter	100	times farther

Getting back to the astronomical use of the inverse square law, why bother with an equation that calculates what you can just step outside and see — a star's apparent brightness? Turn the inverse square law inside out and you'll see why:

$$\text{Luminosity} = \text{apparent brightness} * 4 * \pi * \text{distance}^2$$

Put apparent brightness and distance into the inverse square law and out pops a star's luminosity, one of the inate properties of a star. That is the hidden beauty of the inverse square law. But there is a catch. You can measure a star's apparent brightness very easily, just go outside and look at it. But how do you measure its distance? That, as you'll see, is the fundamental problem in astronomy — measuring the distances to things we cannot touch.

A Sky Full of Angles

The first step to measuring the radii of stars is to see how big they look. An obvious enough statement, but what exactly do you measure when you measure the sizes of objects by eye? For example, two people are looking out of the window of a tall building. One exclaims to the other "Wow! The people down there look like ants!" The other responds "You dummy, those *are* ants. We're only on the *first* floor." In this anecdote you find all you require to measure the sizes of stars: *angular size* and *distance*.

1 arm's length

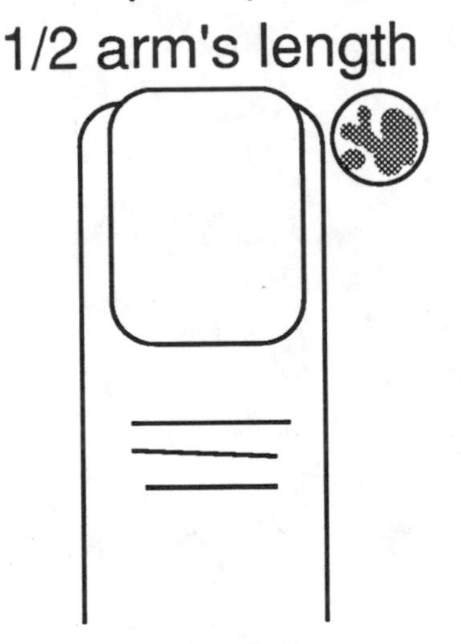

Figure 5-3

To once again restate the obvious everyone knows *perspective*, how objects appear to shrink with distance, intuitively. Just look to the example of 10,000 years of painting. Look back only 500 years ago and you 'll see the large and the small mixed ajumble — kings tower over castles while peasants shrink underfoot. It's not that the artists were poor draughtsmen, rather their work reveals the world as seen from a *social perspective*, where size reflects importance, not distance. Even when artists did attempt *perspective* they failed. They lacked the mathematical tools to turn intuitive knowledge into draftsmanship. They could not draw what they could see. Only in the 15th century was *perspective* finally codified into geometry and the eye's angle expressed as the canvas' inch.

The key to understanding perspective comes from the recognition that the eye sees *angles* not *inches*. Hold your hand out at arms length and your pinky fingernail is as wide as the Moon — in angle, not in inches. When you express sizes of objects in inches you are assuming something about distance and using your intuitive knowledge of geometry to mentally convert your eye's angles to inches. You can explicitly express your intuitive understanding of geometry of perspective with a simple experiment. Once again hold out your hand at arms length so that your pinky fingernail covers the Moon 1/2° wide. Measure the distance from your eye to your fingernail and record that measurement in Table 5-3. Now bring your finger in towards your eye until your pinky fingernail looks 1° across — twice as wide as the Moon. Measure the distance from your eye to your fingernail and again record your results in Table 5-3.

Table 5-3: Size vs Distance

Angular Size °	Distance Inches
0.5	
1.0	

These two measurements roughly reveal the very simple relationship between angle and distance: twice the distance means half the angular size (Figure 5-3):

$$\text{angular size} = 57 * \frac{\text{real size}}{\text{distance}}$$

The 57 in the equation is just there to make the angular size come out in degrees with the real size and distance measured in inches (or miles or kilometers or lightyears), but that's not what's important in the equation. Turn thc law of perspective inside out, just like you did for the inverse square law:

$$\text{real size} = \frac{\text{angular size} * \text{distance}}{57}$$

Put in a star's angular size and its distance, and out pops its real size. Of course measuring angular size for stars isn't all that easy. When was the last time you saw a star as more than a pinpoint? Even the closest is so incredibly distant, some

40,000,000,000,000 kilometers or
25,000,000,000,000 miles

away that it appears as a pinpoint of light to the eye, even with the largest telescope. Worse yet, the law of perspective requires you to put in the star's distance in kilometers if you are ever to convert angular size (degrees) to a real size (kilometers). Once again you have run into the critical problem of astronomy — how do you get the star's distance? Before finally answering that question, there's one more inate property of a star to consider that also requires distance: its mass.

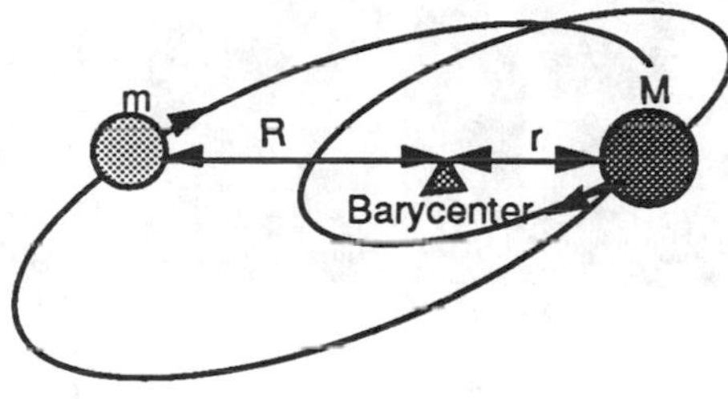

Figure 5-4

Motions and Masses

The way you measure any object's mass, be it a bowling ball or a star, is to wave it around. For example, a ping-pong ball is a lot easier to wave about than a bowling ball. However stars are a bit large and too distant to grab. Instead we rely on stars to pull each other about. Such pairs of stars, called double stars, provide the *only* opportunity to directly measure the masses of stars.

Each star will move in an elliptical orbit about a *barycenter* (the center of mass of the pair of stars, like the balance point of a teeter-totter) in response to the pull of its partner's gravity. The larger mass star staying nearer the *barycenter* while the smaller mass star swings around in a larger orbit (Figure 5-4) or, expressing it mathematically:

$$\text{Large mass} * \text{small distance} = \text{small mass} * \text{large distance}$$

Shift both masses to ones side of the equation and the distances to the other and you find that the ratio of their masses equals the ratio of their distances from the barycenter:

$$\frac{\text{large mass star}}{\text{small mass star}} = \frac{\text{large distance}}{\text{small distance}}$$

To get their individual masses in real units (e.g. kilograms), instead of ratios, requires you to also measure the length of time they take to orbit about each other (their *Period of revolution*) and the sizes of their elliptical orbits in real units (i.e. kilometers). Measuring the *period* isn't too difficult , just start your stop watch and wait for the two stars complete one orbit about each other. Unfortunately, when it comes to measuring the sizes of their orbits, what you observe through the telescope is the *angular* size of their elliptical orbits, not the *real* size of their orbits in kilometers. Of course you can use the law of perspective to convert an angular orbit into a real orbit — provided you know the distance of the star. But once again, how do you find the distance to something you cannot touch?

The Stuff of Stars

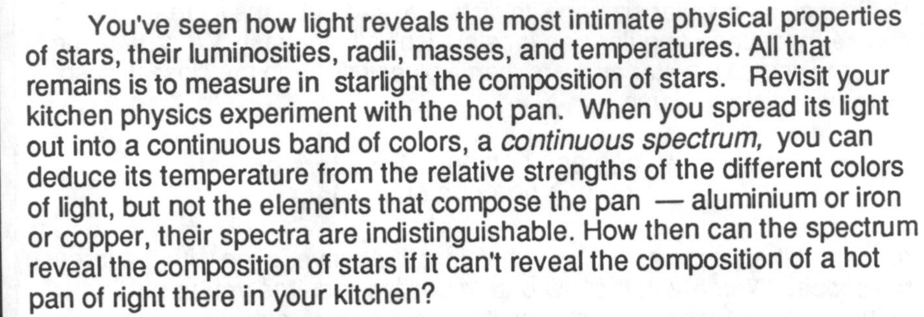

You've seen how light reveals the most intimate physical properties of stars, their luminosities, radii, masses, and temperatures. All that remains is to measure in starlight the composition of stars. Revisit your kitchen physics experiment with the hot pan. When you spread its light out into a continuous band of colors, a *continuous spectrum,* you can deduce its temperature from the relative strengths of the different colors of light, but not the elements that compose the pan — aluminium or iron or copper, their spectra are indistinguishable. How then can the spectrum reveal the composition of stars if it can't reveal the composition of a hot pan of right there in your kitchen?

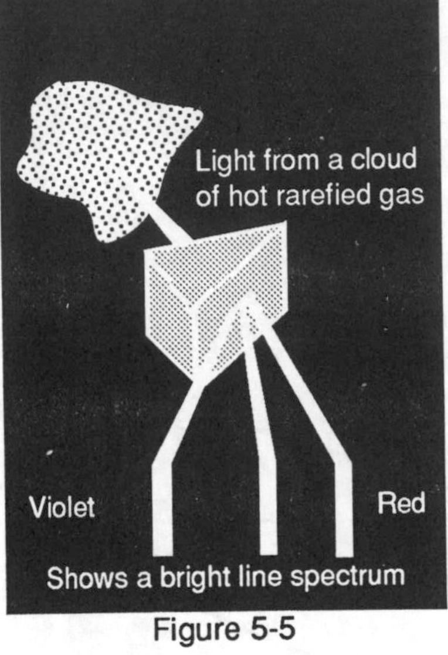

Figure 5-5

The continuous spectrum cannot reveal an object's composition. However there are other types of spectra that can. Heat an *rarefied* (almost a vacuum) elemental gas, say hydrogen, into incandescence and it will produce an entirely different kind of spectrum, a *bright line* spectrum (*emission* spectrum) ; sharply defined colors with darkness in between (Figure 5-5). Heat some other tenuous elemental gas, say iron or aluminum or copper, into incandescence and it too will produce a *bright line* spectrum — but each one with a unique combination of colors of light, of wavelengths. *From element to element the emission line spectrum is unique.*

Just as each atom can emit only certain discrete wavelengths, colors of photons, each atom can absorb only certain discrete wavelengths . Shine light of all colors (*a continuous spectrum*) past a cloud of hydrogen atoms and their electrons will absorb only those photons that have just the right energy (as defined by their wavelengths) to knock the electrons to higher orbits. The spectrum will now show *dark lines* in place of the missing wavelengths (Figure 5-6). The wavelengths of those dark lines in this *absorption* (*dark line*) spectrum exactly match the lines you'd see in the emission line spectrum for the same element. Once again each element reveal its identity by choice of photons — *the spectrum of each element is unique.*

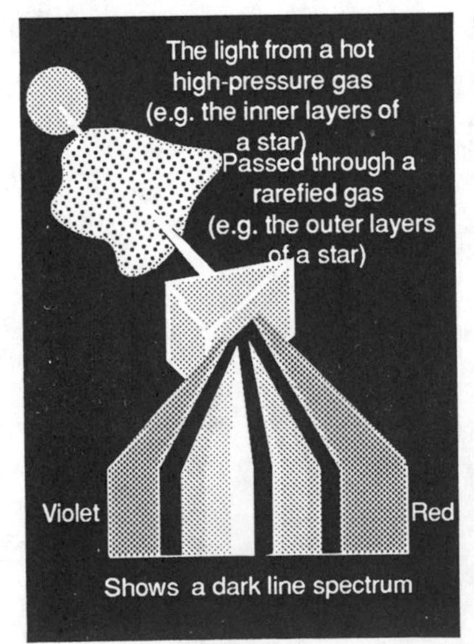

Figure 5-6

Sometimes a continuous spectrum, sometimes a bright line spectrum, sometimes a dark line spectrum, . Why the difference? If all atoms of a particular element in an object indeed had exactly the same orbits for their electrons, then they would all produce the same energies of photons and you would see a bright line spectrum. You can find just such an example in most store windows. Neon signs (which don't always contain neon) use an electric current to heat the *rarefied* (almost a vacuum) gas causing its atoms to emit a bright line spectrum. Because the atoms are well spaced in a rarefied gas they are all pretty much identical to one another and produce identical photons when each atoms electrons make the identical orbit changes.

Pack the atoms more tightly together as in a solid or a liquid and the atoms press into each other distorting each other's electron orbits. The bright line spectra from the aggregate of all the atoms smear out into *a continuous spectrum.* Heat an gas hot enough and the collisions between atoms rip electrons off of one another. These electrons, interact to produce photons of all colors — once again a continuous spectrum. You can summarize these observations of the different kinds of spectra as:

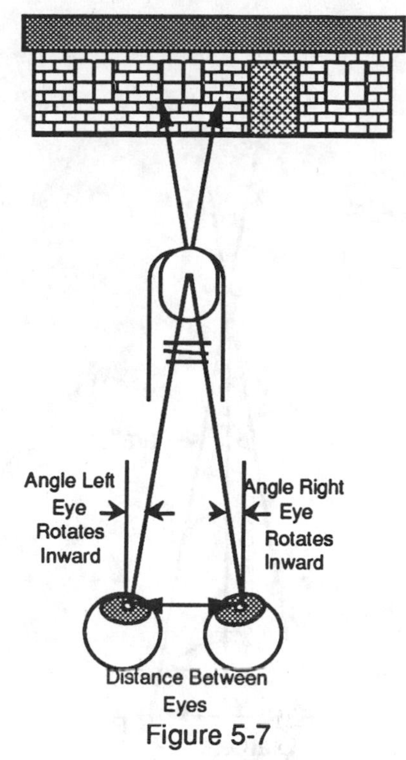

Figure 5-7

1) A Hot solid, liquid or high pressure gas emits a *continuous spectrum.*

2) A hot rarefied gas emits an *emission line spectrum*.

3) If a continuous spectrum from a hot solid, liquid or high pressure gas passes through a cooler rarefied gas, then you'll see an *absorption line spectrum.*

So which of these three types of spectra does a star produce? Look at the rainbow and it looks continuous in color. Look more closely and you'll find fine dark lines across the spectrum. In the absorption spectra of the Sun and stars you'll find the spectral signatures of the elements that compose them to be the same elements that compose the Earth and ourselves.

Probing the Depths of Space

Distance is the philosopher's stone of astronomy. With distance, a star's angular size transforms into radius and its apparent magnitude into luminosity. With distance, the dances of double stars reveal their masses. With distances, we can map the Milky Way and chart the scatter of galaxies back across the universe to the beginning of time. Without distance, we are all just whistling in the infinite dark. Just how do you measure distance to something you cannot touch?

There is one way called *parallax.* You probably know it as depth perception or "3-D" vision. It is how you measure distances to everything in your environment from your nose outward. All it takes is two eyes spaced some distance apart and a little geometry. When you look at a nearby object ,your eyes turn inward (Figure 5-7) ,forming a long thin triangle with the object at one end and your eyes at the other. The nice thing about triangles is that if you know two angles and the distance between them, then you can determine everything else about the triangle. In this case, your brain measures two angles from how far inward your eyeballs rotate and uses the distance between your eyes to solve the triangle for the distance of the object. That's *parallax*, and you perform it at the speed of thought.

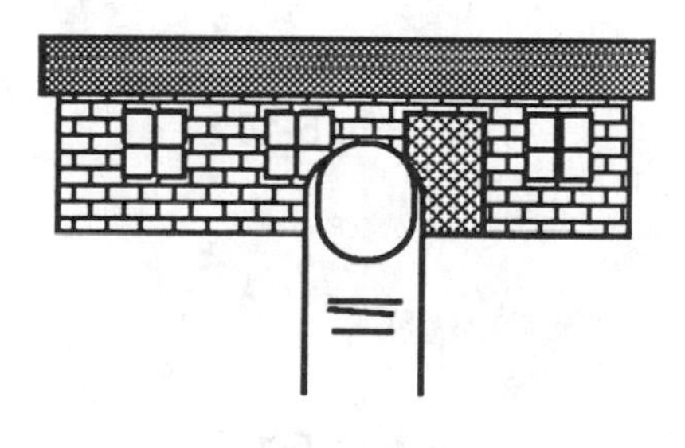
Figure 5-8a: Left Eye

Slow down your thought process and the intuitive becomes logical. Hold out your hand at arms length with your thumb extended upwards. Close your left eye and note your thumb's position relative to objects in the distance. Now close your right eye and open your left eye and you see your thumb jump to the right. Start blinking from one eye to the other and you will see your thumb jumping left and right (Figure 5-8a, 5-8b). Bring your thumb inwards towards your nose and it jumps left and right

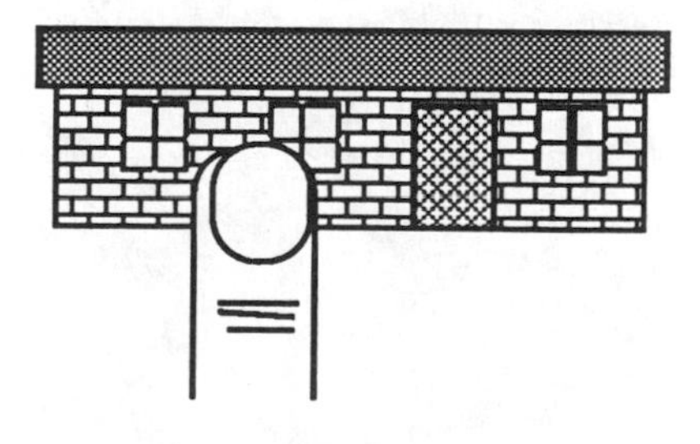
Figure 5-8b: Right Eye

over larger and larger angles. Increase distance and the angular shift decreases. Now just express these observations as a *parallax equation*:

$$\text{Distance} = \frac{1}{\text{parallax}}$$

Parallax is the angular shift you measure. A large *parallax* divided into 1 produces a small distance. A small *parallax* divided into 1 produces a large distance.

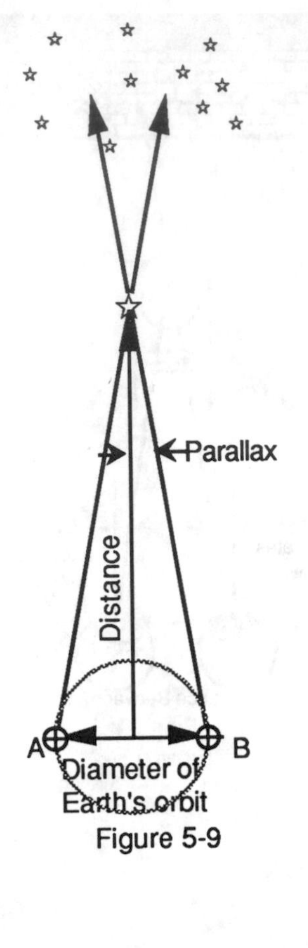

Figure 5-9

Astronomers measure the parallax of stars the same way but with eyes spaced an earth's orbit apart (Figure 5-9). We watch stars over the course of a year as they shift back and forth owing to the earth's changing position as it orbits around the Sun (Figure 5-10a, 5-10b). By convention, astronomers define the *parallax* as half the total angular shift not the whole amount. The *parallax* for even the closest star is miniscule, less than 0.763" (arc seconds), which corresponds to an immense distance. Dividing 1 by 0.763 to get 1.31 isn't impressive in itself, but look at the units. That distance is in *parsecs* and *parsecs* are huge. One *parsec* is:

30,860,000,000,000 kilometers,
19,176,000,000,000 miles

Huge distances from measuring small angles; angles that dwindle to insensibility as you reach further out. Have a friend stand 50 feet away and hold up his/her thumb. As you blink from one eye to the other, you will not see it shift left and right. You can no longer measure the parallax, you can no longer measure distance. Your friend could just as well be a parsec as 50 feet away. The same is true for stars. Reach out far enough and, before we're even out of our galactic backyard, parallax shrinks to insensibility.

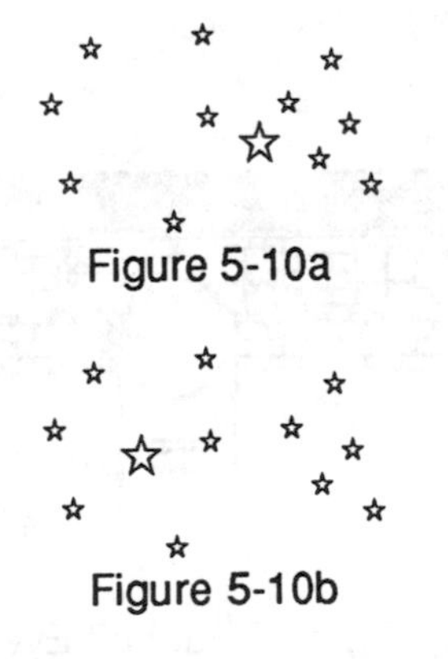
Figure 5-10a

Figure 5-10b

At 20 parsecs, parallax is less than 0.05" and dwindles to the limit of our earthbound ability to measure small angles (0.005") at 200 parsecs. We can expect to extend our reach by 10 times with the Hubble Space Telescope's capability to measure parallax to 10 times smaller. But that is in the future. Here and now parallax dwindles from good to 10% at 20 parsecs to meaningless at 200 parsecs. Perhaps 10% doesn't sound too bad but would you buy a yardstick that was only good to ± 3.6 inches? Apply that 10% error to converting your observations of a star into its physical description and out comes a 10% error in size, a 20% error in luminosity and a 30% error in mass. At 200 parsecs, the errors are astronomical since distance is but a guess. Beyond 200 parsecs, we don't even have a guess with parallax. Beyond 200 parsecs we need some other distance indicators. The only ones available are the inverse square law:

$$\text{apparent brightness} = \frac{\text{luminosity}}{4 * \pi * \text{distance}^2}$$

and the law of perspective:

$$\text{angular size} = 57 * \frac{\text{real size}}{\text{distance}}$$

Turn both inside out and the inverse square law and the law of perspective can be used to measure distance:

$$\text{distance} = \sqrt{\frac{\text{luminosity}}{4 * \pi * \text{apparent brightness}}}$$

$$\text{distance} = 57 * \frac{\text{real size}}{\text{angular size}}$$

Put apparent brightness and luminosity into the inverse square law or angular size and real size into the law of perspective and you can calculate an object's distance. Apparent brightness and angular size are things you can measure just by looking at an object, but where do you get size and luminosity to feed these equations? If only stars had their luminosities printed on top like light bulbs and star clusters came with size tags attached. Well, perhaps they do.

Here is the fundamental assumption required to use the inverse square law and the law of perspective to measure distances. *If two celestial objects look identical in most characteristics, then they are identical in all characteristics.*

For example, if you observe that a remote star has the exact same temperature (from its spectrum) as a nearby star (less than 20 parsecs away) and the exact same lines in its spectrum with exactly the same strengths, then you could also assume that the remote star has the same luminosity as the nearby star that you *can* calculate from its apparent brightness and parallax using the inverse square law. Now use the inverse square law to get the distance to the remote star from its apparent brightness and the luminosity you've assumed for it. As a bonus, if that star lies in a star cluster, then you have measured the distance to that remote star cluster.

If the star cluster resembles an even more remote star cluster , then you might assume that both have the same physical size and use the law of perspective to get the more remote cluster's distance. This process, called *bootstrapping,* is how astronomers measure by extending this process out to nearby galaxies, then remote clusters of galaxies, then super clusters of galaxies, and so on out to the edge of the observable universe.

Unfortunately, bootstrapping is only as good as the chain of assumptions fed into the inverse square law and the law of perspective. Just how reliable is the key assumption that because some celestial objects look identical in most characteristics they are identical in all characteristics? Especially when, for distant objects, we can at best measure only a few of their characteristics given the minute quantity of light we can collect from them even with the largest telescopes?

Suffice it to say that each distance we measure beyond the reach of parallax compounds the errors of all the distance indicators before it: 20% errors on the galactic scale grow to factors of 2 at the largest scale of the universe. We accept these uncertainties in distance only because every luminosity, every mass, every radius, every feature of stars, galaxies, and the universe itself that we wish to measure, every fundamental question of origins and fates hang on the measurement of distance. A distance good to a factor of 2 is better than no distance at all ,for quite simply we have no alternative once we reach out beyond the range of parallax.

Project 5-1
Spectrum, Luminosity, Perspective, and Parallax

Part A: Spectral Analysis

Your Instructor will now excite several elemental gases at the front of the room for your viewing enjoyment. Your viewing will be enhanced by the use of a diffraction grating, a handy optical device which divides the light up into its spectrum of colors. Sketch the spectrum of each element, paying particular attention to:

1) Wavelength (color)
2) Intensity
3) Spacing

of the spectral lines. You can represent wavelength using colored pencils, intensity by the thickness of the line, spacing by the spacing of the lines. Table 1 requests some additional perceptions about the spectra you observe.

Known Element	Spectrum
	Indigo Blue Green Yellow Orange Red
	Indigo Blue Green Yellow Orange Red
	Indigo Blue Green Yellow Orange Red
	Indigo Blue Green Yellow Orange Red

Indigo Blue Green Yellow Orange Red

Indigo Blue Green Yellow Orange Red

Table 1

Element	Number of Lines	Color of Strongest Line	Overall color (without diffraction grating)

You now have the opportunity to apply your knowledge to real world examples. Your Instructor will assign you several common sources of hot gas for your investigation (e.g. street lights, neon signs). For each you will sketch the spectrum and identify the element contained within the source by comparison with the spectra you drew for known elements.

Unknown Element	Spectrum

Indigo Blue Green Yellow Orange Red

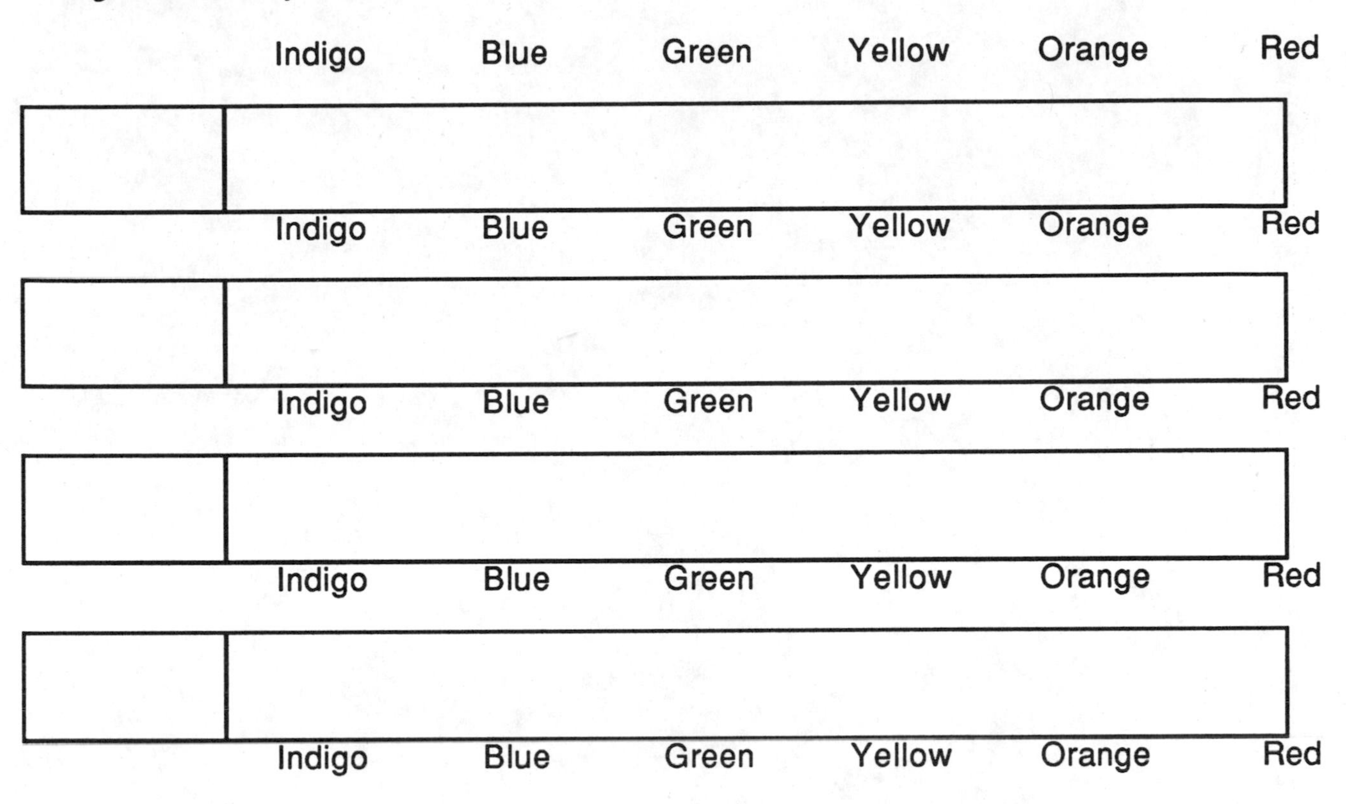

Hints

1) *An element will always produce the same spectral lines with the same spacings but not necessarily with the same intensities under all conditions. If the relative intensities of the spectral lines are different from one case to another, then the overall color (the color of the gas, without a diffraction grating) will be different.*

Questions

1) From your observations of the known element spectra, how do the colors and strengths of the spectral lines dictate the overall color (observed without a diffraction grating)? Describe how you can use the strengths of the spectral lines to predict the overall color.

2) What were the unknown elements? Could you identify all the elements?

3) Some of the unknown elements were contained in colored glass tubes. What spectral properties did you use to identify them correctly: overall color (without a diffraction grating), strength of strongest line, wavelengths, and spacing of spectral lines?

4) Did any of the unknowns have an absorption line spectrum? What does that tell you about its physical state? What was the element?

Part B: Inverse Square Law

Your instructor has prepared a number of lamps preset to different luminosities. Using the inverse square law, you will measure their relative luminosities.

1) Find a large darkened area and place the faintest of the lamps about 3 meters away. This is your standard for comparison to the others.

2) Have your partner carry another of the lamps further away until, to your eye, it appears the same visual magnitude as the standard at **3** meters.

3) Measure the distance to your partner and record it in Table 2.

4) Calculate the relative luminosity of the distant lamp to the standard at 3 meters and record your answer in Table 2.

   Hint: Since you just want the luminosity of the distant lamp relative to the luminosity of the standard lamp, you divide the inverse square law for the distant lamp by the inverse square law of the standard lamp:

$$\frac{\textit{luminosity of distant lamp}}{\textit{luminosity of standard lamp}} = \frac{\textit{distance}^2}{(\textit{3 meters})^2}$$

   Presume the standard lamp's luminosity as 1, and the equation simplifies to:

$$\textit{luminosity of distant lamp} = \frac{\textit{distance}^2}{9}$$

5) Repeat steps 2 through 4 for the other lamps.

6) When all is said and done, obtain the actual luminosities from your instructor, record them in Table 2, and calculate your luminosity error by subtracting them from your calculated luminosity.

Questions

5) Compare your results to the correct values from your instructor. Were you close (within 10%)? Did you systematically over- or underestimate luminosity at short or long distances? Describe.

   Hint: Plot your values versus the correct ones on a graph.

6) Were there any detectable color differences among the lamps? Why might you expect there to be?

   Hint: Since all the grain-of-wheat bulbs are the same size (have the same surface area), what property makes the more luminous lamps more luminous?

Table 2

Lamp	Distance meters	Calculated Luminosity	Actual Luminosity	Luminosity Error
Standard	9	1	1	0

Part C: Perspective

1) Have a partner stand 3 meters away and measure the angular size of his/her head in multiples/fractions of your pinky finger nail (half a degree) and record results in Table 2.

   Hint: You may want to use a ruler rather than try to count fingernails. So measure the width of your pinky finger nail in millimeters, double that number since your pinky fingernail is half a degree and you have the number of millimeters corresponding to a degree when you hold the ruler up at arm's length. Now you can hold the ruler up at arms length to measure sizes in millimeters and convert that measurement to degrees.

2) Have your partner move another 3 meters away and repeat steps 1 and 2 until your partner is 30 meters away.

3) Use your measurement at 3 meters and the perspective equation to calculate distances to your partner:

$$\text{distance} = 3 \text{ meters} * \frac{\text{angular size}}{\text{angular size at 3 meters}}$$

4) Calculate the error in your distances by subtracting them from the actual distances.

Questions

7) Does angular size get larger or smaller with distance? How rapidly does size change with distance?

   Hint: Plot the angular size of your partner's head versus his/her distance on a graph. Does angular size drop evenly with distance (twice as far, twice as small) or at a faster rate (e.g., inverse square — twice as far looks four times smaller — or inverse cube — twice as far looks 8 times smaller).

8) How well do your measurements agree with the laws of perspective? Do you find yourself over- or under-estimating angular size at short or large distances? Describe.

   Hint: Plot your calculated distance versus actual distance on a graph and describe what it shows.

Table 3

Actual Distance meters	Measured Angular Size °	Calculated Distance meters	Distance Error meters
3	3	0	
6			
9			
12			
15			
18			
21			
24			
27			
30			

Part D: Parallax

Parallax is the *only* direct way to measure distance to stars. The technique is intuitive, its how you get distance to the world around you. Your eyes pivot inwards as you focus on an object. Your brain measures the distance they pivot inwards and since it knows the distance between your eyes, it calculates the distance to the object.

1) Using your parallacter (see next page), measure the parallax of objects at the distances listed in Table 4.

2) Convert the parallaxes you measure to distance using the value you measured for your eye spacing in the following equation:

$$\text{Distance in millimeters} = 57 * \frac{\text{your eye spacing in millimeters}}{\text{parallax}}$$

3) Calculate the error in your distances by subtracting them from the actual distances.

Table 4

Actual Distance meters	Parallax °	Calculated Distance millimeters	Distance Error millimeters
1			
1.5			
2			
2.5			
3			
3.5			
4			
4.5			
6			
6.5			
7			

Questions

9) Does parallax get larger or smaller with distance? How fast does it change with distance? What do you think could contribute to your errors?

   Hint: Plot your measurements of parallax versus distance on a graph and describe what you see.

10) Do your measurements of parallax lead to over- or under-estimation of distance for small parallaxes? At what distance does parallax produce meaningless results?

   Hint: Plot your calculated distance versus actual distance on a graph and describe what you see.

Assembly

1) Cut out parallacter along solid black line and glue to heavy cardboard. Cut out the nose notch.

3) Measure the distance between your eyes by holding a ruler up to your eyes as you look into a mirror. Close your right eye and line up the zero end of the ruler with your left eye's pupil. Now, without moving the ruler, close your left eye, open your right eye and read off the measurement of the center of your right pupil. The answer will be somewhere between 55 and 75 millimeters.

4) Tape a soda straw between A and B along the line corresponding to your eye spacing. Make sure its attached tightly and doesn't move. This is your fixed sight.

5) Align another straw between point C and the degrees scale but tape it only at point C so that it can pivot slightly. This is your moveable sight.

6) Trim the end of the moveable sight at the degree scale (the numbers zero to 8).

Operating Instructions

1) Hold parallacter up to your nose and look through fixed sight with your left eye at object.

2) While continuing to look through fixed sight with your left eye, pivot front of moveable sight until you see the same view through it with your right eye. Align both views in your head.

3) Hold moveable sight steady, take parallacter off your nose and read off the number (0-8) aligned with center line of moveable sight.

4) This is the parallactic angle.

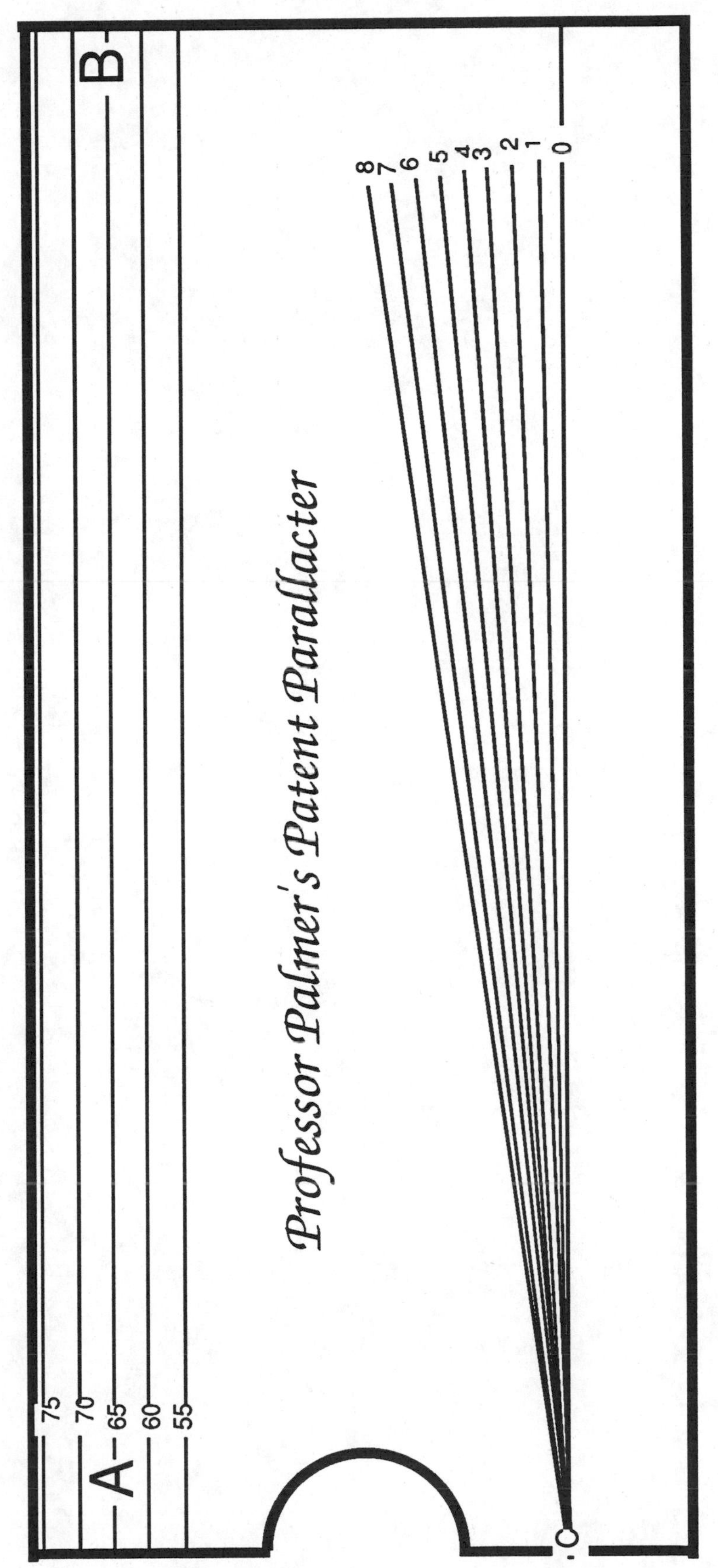

Chapter 6
Reading the Message in Starlight

Mass, luminosity, radius, temperature, and chemical composition provide the vocabulary of the message in starlight — the measure of any individual star (Table 6-1). Yet, important as these descriptions are for any one star, what simple theme unites these disparate descriptions — what common story do stars tell? What do you learn about stars in general? After all, that is the primary assumption of astronomy and all other sciences; the complexities we observe in nature are aspects of a few simple common truths. This belief drove Kepler to discover the simple celestial harmony in the motions of the planets from Tycho's complex observations. The same belief drove Newton to unite terrestrial and celestial physics into a universal physics.

Table 6-1: Properties of Stars

Star	Mass M_{sun}	Luminosity L_{sun}	Radius R_{sun}	Temperature °Kelvin
Sun	1	1	1	5770
σ Aql A	6.8	460	4.2	12,000
σ Aql B	5.4	180	3.3	11,000
δ Lib A	2.6	170	3.5	9,900
δ Lib B	1.1	10	3.5	5,300
β Per A	5.2	200	3.6	12,000
β Per B	1.0	6.6	3.8	4,500
λ Tau A	2.3	1500	3.4	22,000
λ Tau B	0.9	180	4.8	9,000
η Cas A	0.8	3.3	0.84	6,000
η Cas B	0.5	0.03	0.07	3,500
α CMa A	2.3	210	1.8	9,700
α CMa B	1.0	0.002	0.022	10,000
α CMi A	1.8	7	1.7	6,580
α CMi B	0.6	0.0005	0.01	6,600

You can sift through Table 6-1, even expand it a hundredfold, looking for a common theme among the properties of stars and still not find it. Not because it isn't there; it is. It's simply because tables obscure answers. Take the data from the table and convert it into a picture and the simple relationships among the properties of stars leap out. But just how do you convert a table full of numbers into a meaningful picture?

The appropriate type of picture is a graph where you plot the numbers associated with one stellar parameter against the numbers associated with another stellar parameter for each star. If the stars scatter all over the graph, then the two stellar parameters you've plotted are independent — there is no common theme (Figure 6-1a). If the stars cluster together on the graph, then the two stellar parameters are dependent; united by a common cause (Figure 6-1b). When you find such dependencies, you've glimpsed simplicity among complexity and the harmony in nature.

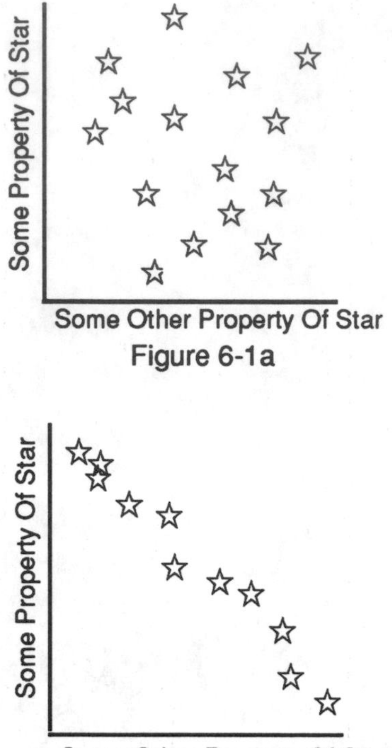

Figure 6-1a

Figure 6-1b

Which pair of parameters should you start with from Table 6-1? There are, after all, six different ways to pair off mass, luminosity, radius, and temperature. But there is only one way that captures the essence of stars. That graph is called the Hertzsprung-Russell (abbreviated H-R) diagram after Hertzsprung (1911) and Russell (1913), who independently created the first versions of the H-R diagram by plotting luminosity versus temperature for a number of stars (Figure 6-2).

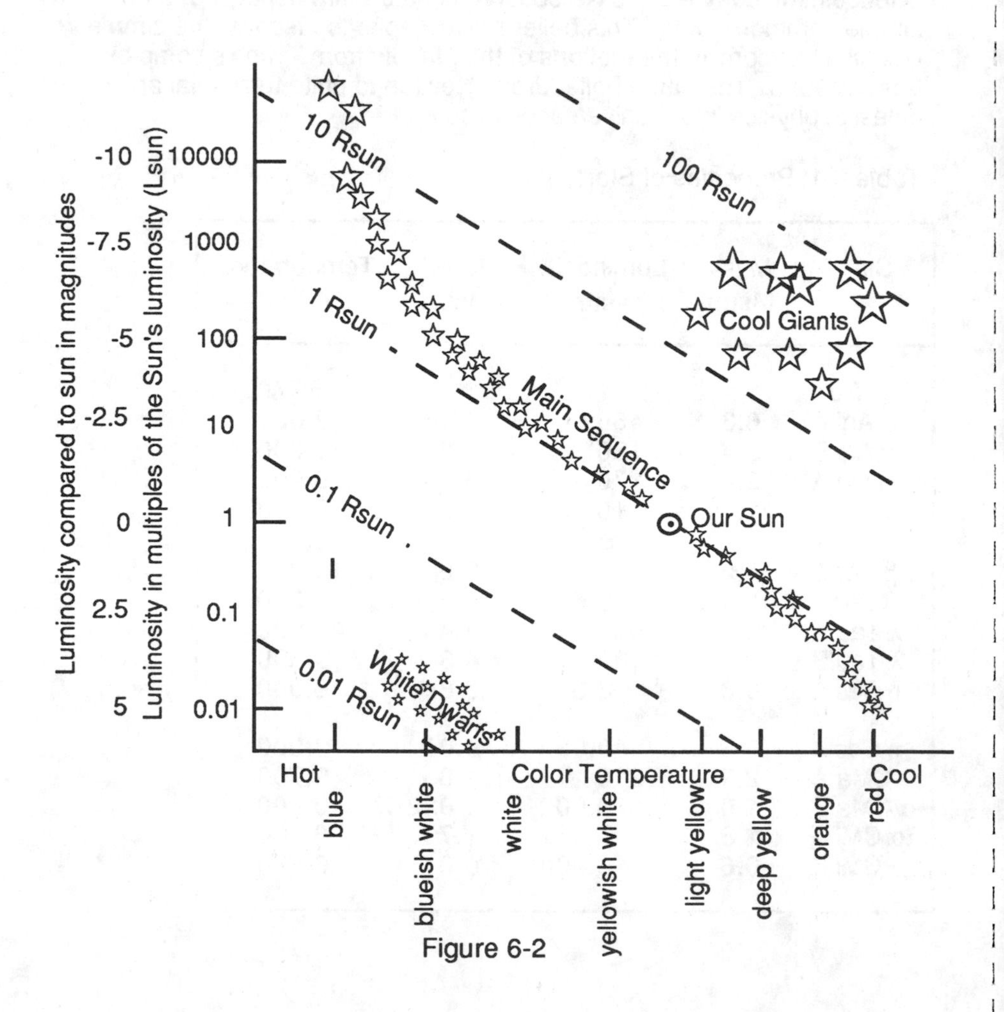

Figure 6-2

Look at this H-R diagram and you see that the stars do not fall randomly about the diagram — 90% of the stars fall into a band slanting from the upper-left corner (hot, bright) to the lower-right (cool, faint), which Russell called the *main sequence*. Most of the remaining 10% either fall into the upper right (cool, bright — *cool giants*) corner or into the lower left (hot, faint — *white dwarfs*) corner, the rest scattering about with some areas having almost none.

Note the implication of *cool giant* and *white dwarf*; the relative locations of stars on the H-R diagram immediately reveals their relative sizes! That should come as no suprise. As you know, a star's luminosity reveals the total flux of photons it emits while its surface temperature defines its *surface brightness*, the flux of photons emitted per unit area. Since you know the total flux of photons (luminosity) and the flux emitted per square inch (surface brightness), then you can calculate its radius. This has already been done for you in Figure 6-2, where the sizes of stars are shown as diagonal lines sloping from upper right to lower right. Follow the 1 R_{sun} line down to where it crosses the main-sequence and you find our Sun (for convenience, we use the Sun as our standard unit and measure star radii in units of the sun's radius, luminosity in units of the sun's luminosity, and mass in units of the sun's mass and label them R_{sun}, L_{sun}, and M_{sun}, respectively). White dwarfs huddle down around 1/100th R_{sun} while cool giants range between 10 to 100 R_{sun}. With size now revealed upon the H-R diagram, all that remains is to add mass.

So where on the H-R diagram might we find stars of different mass? You already see three groups: cool giants, main sequence, white dwarfs. Do each of these correspond to a different mass of star or does each group contain stars of all masses? To make mass' relationship to the H-R diagram clearer, you need to create one additional graph — mass versus luminosity (Figure 6-3).

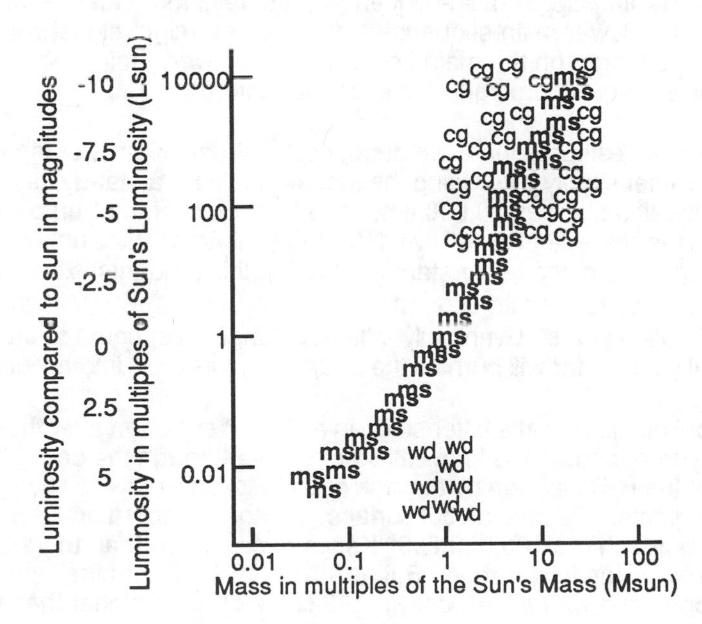

Figure 6-3

Plot the masses for cool giant, main sequence, and white dwarf stars versus their luminosities and the relationships reveal themselves quite quickly. Cool giants don't show any clear relationship — any two cool giants sitting side by side on the H-R diagram, having identical temperatures and luminosities, can have quite different masses. White dwarfs, at the other extreme, all cluster around one 1 M_{sun}. main sequence stars show a definite progression between their masses and luminosities; the greater the mass of a main-sequence star, the greater its luminosity. There is a clear harmony among mass, luminosity, radius, and temperature. Such harmonies demand explanation, why should there be such simple relationships between disparate parameters, what are the simple truths that underlie them? Kepler's discovery of the harmony among the motions of the planets was just such an empirical discovery, derived much the way you found the relationship between mass, luminosity, radius, and temperature from the H-R diagram.

Kepler's empirical relationships set the stage for Newton to demonstrate the universality of physics; to discover the explanation. The H-R diagram does the same for *astrophysics* (the physics of astronomical objects): the simple relationships among mass, luminosity, temperature, and radius for main-sequence stars led to the discovery of nuclear fusion, the peculiarities of the white dwarfs led to the recognition of new states of matter, the generalities of cool giants reveal the fate of our Sun, and all show gravity as the driving force. The H-R diagram, as you see, is more than just a graph of numbers; it is the stage upon which stars act out their life stories, a story told by mass, luminosity, radius, temperature, and chemical composition.

The main sequence reveals the points on the H-R diagram where stars of different masses first achieve the central temperatures required for hydrogen fusion. High-mass stars live at the upper left (hot, luminous), low-mass stars at the lower right (cool, faint), with stars of intermediate mass in between. How long a star rests on the main sequence depends upon two things: how fast it burns its hydrogen fuel and how much fuel it has to start with. Upper main-sequence stars have more fuel — more mass — than lower main-sequence stars, and you might at first expect them to live longer on the main sequence. However, their prodigious luminosities require prodigious fuel consumption.

You can estimate the main-sequence lifetimes of stars compared to the Sun rather simply by dividing the fuel available for a star by its luminosity. If a star has 10,000 times the luminosity of the Sun but only 10 times the mass, then it will live 1/1000th as long as the Sun. For hot, bright upper main-sequence stars (50 M_{sun}), this amounts to few million years, for cool, faint main-sequence stars (1/10th M_{sun}, 1/1000th L_{sun}), to 1,000 billion years. Eventually all good things must come to an end. Eventually each star will burn all the hydrogen in its core into helium.

Cool giants reveal a latter stage in all stars' evolution after they run out of hydrogen fuel. We know all stars evolve through the cool giant region of the H-R diagram by the lack of a mass-luminosity relationship for the cool giants. Despite a cool surface, the core temperatures for cool giants approach the 100,000,000° Kelvin temperature that triggers the burning of helium into carbon. But helium is an inefficient fuel compared to hydrogen and with a luminosity a 100 times or more higher than while on the main sequence, helium fusion soon plays out.

White dwarfs reveal a final resting state for low mass stars that have exhausted their nuclear fuels. Crushed by gravity into dense balls of carbon and oxygen hardly larger than the Earth, they slowly fade into eternity huddled in the lower left corner of the H-R diagram.

White dwarf, cool giant, or main sequence — the H-R diagram, reveals their interdependence and the interdependence of luminosity, temperature, mass, and radii . Looking at a star's position on the H-R diagram you know its current phase of life, how long it can stay in that phase, what precedes and what follows. Mass, luminosity, radius and temperature reveal through the H-R diagram the life stories of stars, *the message in starlight*; Stars are faraway suns, massive spheres of hydrogen infused with fire by gravity — damned at birth, for gravity consumes her children. Still they rail against her embrace with the light of their incandescence. Luminous exclamations that fade, across time and space, to faint twinkly lights in the darkness.

Beauty comes with understanding. As you've applied the H-R diagram to individual stars, you can also apply the H-R diagram to understanding your observation of other celestial objects. For, truth is, there are more things in the heavens than stars and planets. An obvious thought now, but once unforseen.

Multiple Star Systems

The Trained Eye Star Atlas lists a number of *multiple star systems* and descriptive parameters that you can use to distinguish their differences and confirm their similarities: *visual magnitude, separation, position angle*. *Visual magnitude* is given separately for each component star of the multiple star. Component stars are named according to their brightness. The brightest star is *A*, second brightest *B*, etc. For example, Sirius (α CMa) of Egyptian fame consists of two stars: Sirius A, a bright shining gem, and lost in its glare, its companion star — Sirius B, some ten magnitude fainter. *Separation* (the distances between each pair of component stars in the multiple star system, measured in seconds of arc — Figure 6-4) defines a size dimension for the grouping of stars in a multiple star. Seperation alone isn't sufficient to define completely the grouping of the components in a *multiple star system*. You also need a parameter to describe their relative orientations. That parameter is *position angle* (the orientation of the stars on the celestial sphere measured clockwise from north).

Just because these parameters are printed in The Trained Eye Star Atlas doesn't mean you will see a multiple star system as multiple stars. As you can well expect (from Chapter 3), your telescope sets limitations on their *visibility*. You can use the apparent visual magnitudes, separations and position angles for the components of a multiple star system to judge whether you can see the components of a multiple star system individually through your telescope. The closer the component stars are in visual magnitude and the further apart they are in separation, the easier they are to see. Well, this does, after all, make sense. You can't see a candle next to a searchlight, and if two stars are too close together, their blur disks overlap (Chapter 3) — they aren't resolved. Even when a multiple star system does have bright, well separated components, you may not see them individually until you use more magnification.

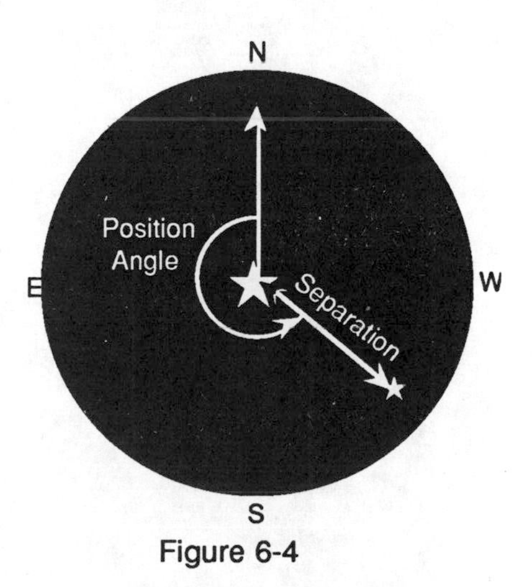

Figure 6-4

How do you select the right eyepiece to get the right magnification? Pick an eyepiece with a focal length equal to or shorter than the separation of the stars. This eyepiece selection rule assumes an f-ratio of 6 to 8 and objective diameters of 6 to 10 inches (typical of most small amateur telescopes) and novice eyes. If the stars in a double star system are 20" apart you need a 20 mm focal length eyepiece to just see both stars as individuals. Want to see them better? Use a 15 mm or a 10 mm eyepiece. You'll soon find your own rule from experience with your telescope and your eyes.

Regardless of whose rule you use, you'll have to bend it for multiple star systems with *unequal components* — component stars widely different in visual magnitude. For such combinations, you'll need even higher magnification (shorter focal length eyepiece) to move the candle out of the glare of the searchlight. These techniques will help develop your observing skills upon multiple star systems, but beyond that why do you want to bother with multiple star systems at all?

You bother with them for their beauty; for their striking appearance — jewel-like colors set against black velvet. Their *descriptions* in The Trained Eye Star Atlas only hint at their beauty. Physical beauty isn't the only reason — you can see pretty colored lights on any street during the holiday season. You must also consider multiple star systems' philosophical beauty, a beauty inherent in the discoveries of Sir Isaac Newton. Sir Isaac Newton, the sage of gravitation, declared his laws of physics and gravitation *universal*. Quite a display of chutzpah, for he had no evidence of their operation beyond the solar system. His was a leap of faith — faith that the universe followed one set of physical laws. Newton's assertion refuted the accepted reality of his time; the Earth corrupt, changeable and mortal — the heavens perfect, immutable and divine (a view proposed 2000 years before by the pagan Aristotle and absorbed into Christian dogma). It is this display of universality of physics that captures our eyes and minds.

It wasn't until 1803, when Sir William Herschel proved that components of a multiple star system orbited about each other, that Newton's bold assertion was tested. Newton was correct. The laws of physics are universal. The Earth and the heavens share the same divinity.

Box 6-1: Perception and Observation

An *observation* is more than just a sketch or description of what you see. It is the totality of information that allows someone else (e.g. your instructor) to confirm independently what you saw. An *observation* includes the following categories or types of information identified on your observing sheets (e.g., Figure 6-5):

The items under each category are shown in the example observing sheet (Figure 6-5). Most of the items are self-explanatory except for the following;

Object Uniquely identify what you observed.

Dimension How big is it. For a multiple star this is the separation of the component stars.

OBJECT
Name α Bet RA 24^h 00^m
TypeDouble Star Dec 89° 00'
Vis Mag6.78, 8.76 Epoch 2000
Dimension 30" sep

TELESCOPE
TypeNewtonian Eyepiece 20 mm
Aperture203 mm Magnif 81x
F-Ratio 8 Filter none
Focal Len1626 mm

ENVIRONMENT
WhereDark Sky, Ca Latitude 35° N
DateFeb 29, 1999 Longitude 120° W
Time12:00 pm Altitude 60°
Seeing 1 Azimuth 180°
Transp clear Conditions new moon, light wind
Visibility 2

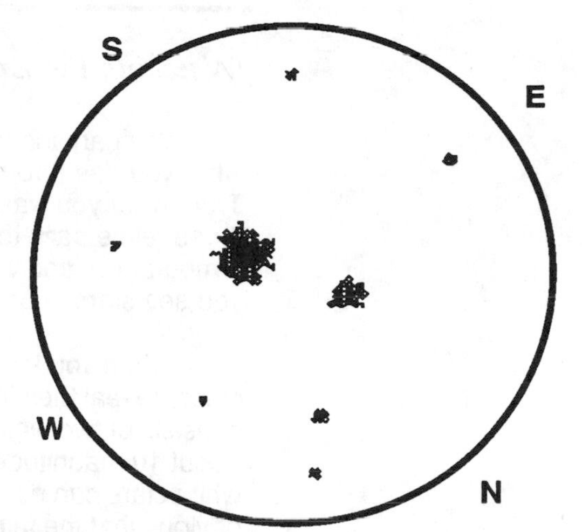

NOTES
Spectacular! Both stars ruby red in color. Fainter star well separated from light of primary. Four other stars in view make an interesting pattern.

Figure 6-5

Epoch	The origin of celestial coordinates drifts slowly through the stars due to precession. As a result the celestial coordinates of a star change slowly. The epoch specifies the date for when the celestial coordinates are exactly correct.
Telescope	Uniquely identify what observing equipment you used.
Filter	Any filters used to enhance viewing. This may include colored filters to bring out planetary surface details or filters designed to suppress light pollution.
Environment	Your terrestrial observing situation and conditions.
Longitude Latitude	Exact location of where you are when you're out in the middle of nowhere (e.g., the Mojave desert) or somewhere (e.g., Los Angeles).
Seeing	A measure from 1 to 6 of the size of star images as they vary from crystal clarity to a soft blur. If the air is steady ,they appear as pinpoints limited only by diffraction. If the air is turbulent, they appear as big fuzzy blurs.
Transparency	How clear is the atmosphere, from crystal clear to socked in.
Conditions	Level of light pollution, light contribution of moon, weather conditions, etc.
Visibility	A measure of how easy the object was to see, from 1 (easy) to 6 (difficult).

Sketch	Worth a thousand words.
Notes	Your explanation and expansion on details in sketch and any other useful comments.

Watching the Evolution of Stars

You can and should apply your knowledge of the H-R diagram to what you see through the telescope, to visual magnitude and color. True, what you want is luminosity and temperature, but you know how those relate back to visual magnitude and color. Color directly tells you temperature, and visual magnitude will suffice in place of luminosity when you see stars in groups.

When you look at a double star, their *relative* visual magnitudes and colors reveal their locations on the H-R diagram. For example, Sirius consists of two white stars (Sirius A and Sirius B) differing in luminosity by about 10 magnitudes. Place them both on the H-R diagram where two white stars can be separated by 10 magnitudes (Figure 6-4) and it is obvious that the fainter must be a white dwarf and the brighter a warm main-sequence star. Much more luminous than the Sun and destined for a short life. The white dwarf, once more but now less massive than its companion now on the main sequence, has evolved through red giant and planetary nebula phases. The white dwarf presages the future for its brighter companion. It too will end the millennium as a white dwarf 100 times smaller and 10,000 times fainter.

Find a pair with a bright yellow component and a fainter blue component and the two differing by a few magnitudes, like γ Andromedae A and B, and you foresee a different ending for the two

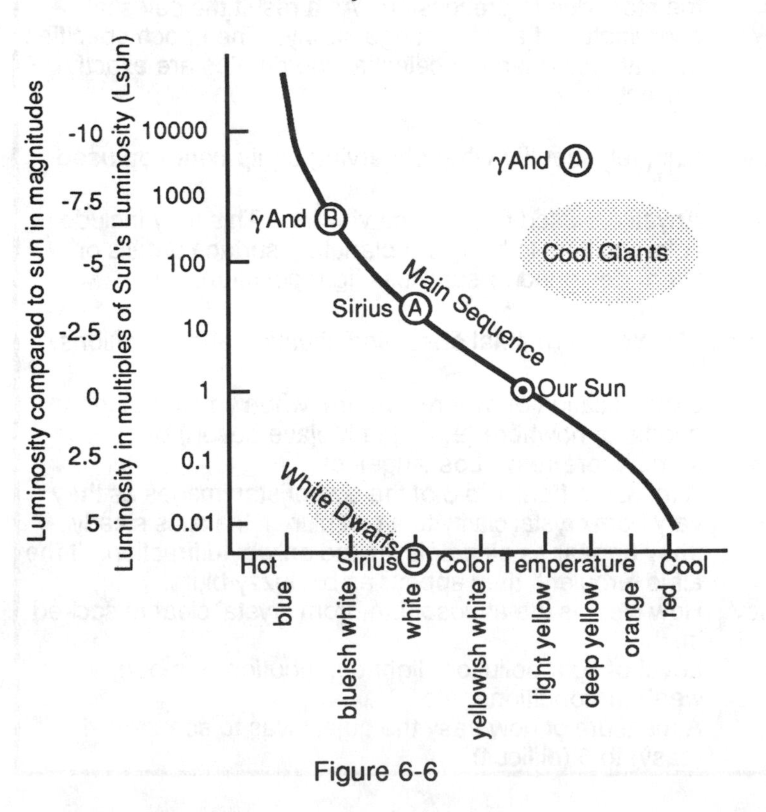

Figure 6-6

stars. γ Andromedae consists of a yellow star with a blue companion star 3 magnitudes fainter. Place both on the H-R diagram where you would expect to find yellow and blue stars separated by 3 magnitudes (Figure 6-6) and you find a yellow giant, already evolved off of the main sequence, and its companion a hot main-sequence star. Since you know that blue main-sequence stars, being so much more luminous than the Sun, do not survive for more than a few 100 million years, you know both component stars in γ Andromedae are a few 100 million years old. You also know that the yellow giant was once an even hotter, more massive main-sequence star than its blue partner since it evolved off of the main sequence first. You also know it to be destined for greatness: a supernova and then a neutron star. By visualizing the H-R diagram, *taking the measure of stars*, you bring life to the view through the telescope — *you understand the message in starlight.*

Box 6-2 Perception and Color

As related in Chapter 2, color vision fades with the dimming of light. In the twilight region of color vision, perception depends as much upon the observer as the stars, for perception of color involves the mind as well as the eye:

1) Color is harder to perceive for point of light than for extended lights. You'll find it advantageous to defocus bright stars slightly to bring out their colors.

2) Faint stars will look bluer than bright stars of the same temperature (the Purkinje shift). Night vision, supplied by your rods cells, is more blue sensitive than normal color vision and almost totally insensitive to red light. Colors of objects (including stars) are a comparison of the relative perceived strengths of different wavelengths; if you don't perceive red light in the object, then it will look bluish.

3) Visual magnitude estimation depends very strongly upon color for faint stars since the eye's sensitivity to different colors changes from color vision to night vision (Dr. Purkinje strikes again). A faint blue star will look brighter than a faint red star of the same visual magnitude because your night vision doesn't see the red light from the red star.

4) The eye excels at comparisons. You can more easily perceive the color of a faint star that is surrounded by stars of contrasting colors than a faint star alone in the eyepiece.

5) Hues of the same color are easier to tell apart at certain wavelengths than at others. The sensitivities of your three types of color cells overlap somewhat in wavelength sensitivity. Where they overlap, you can perceive small changes in hue more readily.

6) Expectations affect perception. If you know the color of the star you will perceive that tinge to its light. Lemons look yellow even when you wear rose-colored glasses. The mind is inextricably involved in vision.

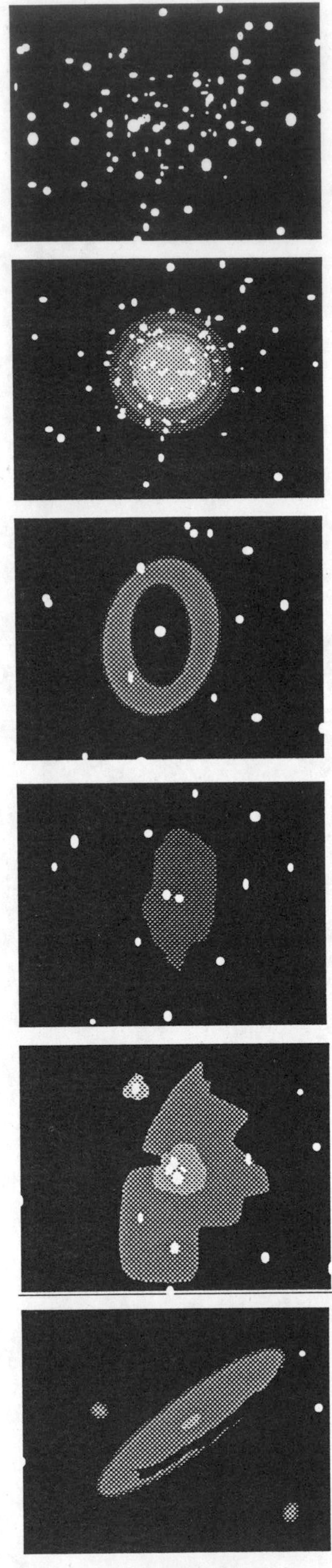

Figure 6-7

7) The atmosphere affects star colors as easily revealed in the change from the ruddy glow of the rising Sun to the white glare of high noon. The reddening of starlight occurs most strikingly below altitudes of 30°. You can avoid most of the effect by observing at altitudes above 30°.

Seeing color is tricky. Train your low-level color vision by progressively defocusing bright stars of known color. Since the star's temperature doesn't change as you defocus its image — it remains the same color — the changes in its hue you perceive as its image spreads and fades reflect your changing perception of color with brightness. With a little practice, you can train your eye to perceive the colors of stars at low-light level — to read and comprehend the color component of the message in starlight.

Non-Stellar Objects

Edmund Halley, a contemporary of Newton, proposed to explain the motions of comets as a test for the universality of Newton's physics. Halley propounded a single comet with a 76-year orbit to explain historical sightings of a bright comet every 76 years. If he was correct then he could predict the comet's next apparition. The French astronomer *Charles Messier* found Halley's comet (as it was thenceforth known) on Christmas day 1787 within a few days of Halley's prediction. Thus was Halley's immortality secured, Newton's physics made triumphant, and Messier inspired.

Messier, comet hunter par excellence, went on to discover and name several other comets, since forgotten. Messier achieved immortality for a class of objects he at first wished to avoid: objects not to be confused with comets. Comets do not burst full-formed into visibility against the heavens. Rather they first appear as faint fuzzy patches drifting slowly against the background of stars, brightening slowly over months. Messier became increasingly distracted in his pursuit of comets by faint fuzzy patches that didn't move. They obviously weren't stars because they didn't look like stars and they obviously weren't comets since they didn't move. Frustrated, Messier tabulated these *non-stellar* objects and published his list as an advisory to other comet hunters. At some point the truth of what he had discovered came to Messier; these non-stellar objects were indeed something new in the heavens, harbingers of new cosmic truths, worthy of study in their own right.

Inspired by this insight, Messier expanded his search for non-stellar objects in a race with other astronomers to see who could discover the most. Messier eventually compiled some 103 objects into his list. Today we still refer to these 103 non-stellar objects by their Messier *(M)* numbers in tribute to him. Still, Messier never pretended to execute a systematic search for non-stellar objects. That was left for J. L. E. Dreher who compiled over 13,000 non-stellar objects into his *New General Catalog (NGC)* and its extension the *Index Catalog (IC)*. As you look through the non-stellar object lists in The Trained Eye Star Atlas, you'll find objects referred to by their NGC, IC, or M number. The thirteenth object on Messier's list, M13, was cataloged by Dreher as NGC 6205, but cataloging doesn't answer the basic question, what *is* M13?

We now know M13 to be a *globular cluster*, a condensed cluster of over a half million stars. In fact, the only common trait non-stellar objects share is their non-stellar appearance in small telescopes. We now know that non-stellar objects divide into open star clusters, globular star clusters, planetary nebulae, supernova remmnants, massive clouds of ionized hydrogen gas (H II regions), and galaxies (Figure 6-7, top to bottom). Just as for multiple stars, each non-stellar object possesses a physical and philosophical beauty; each provides a special perspective on the nature of the universe and all are intimately intertwined in the fabric of the universe. HII regions, planetary nebulae, and supernovae reveal themselves as masses of glowing gas by the colors of their emission line spectra, excited to emission by the flux of ultra-violet light emitted by the stars they contain. Yet they are so different: HII regions are the swaddling cloth of stars just born, planetary nebulae and supernovae are the burial sheets of others. The Trained Eye Star Atlas, under descriptions, summarizes the nature and appearance of each as an aid to finding them. The more you learn about each type of object (from other readings), the more you apply understanding to the observation, the closer you'll come to true observing.

Star clusters provide a good example. Place the stars from a cluster onto the H-R diagram according to their magnitudes and colors (Figure 6-8), and each cluster reveals its age by the stars just leaving the main sequence for the cool giant phase of their evolution. The higher up the main sequence this *turn off point* lies, the younger the star cluster. The turn off point also tells you the masses of the stars leaving the main sequence and their eventual fate as white dwarfs, neutron stars, or black holes. For many clusters, the trail to the red giant phase is plainly marked by the straggle of stars from the main sequence turn-off point. It doesn't take detailed measurements of magnitude and color to see the ages; the ages are plain to see on the H-R diagram in the colors and

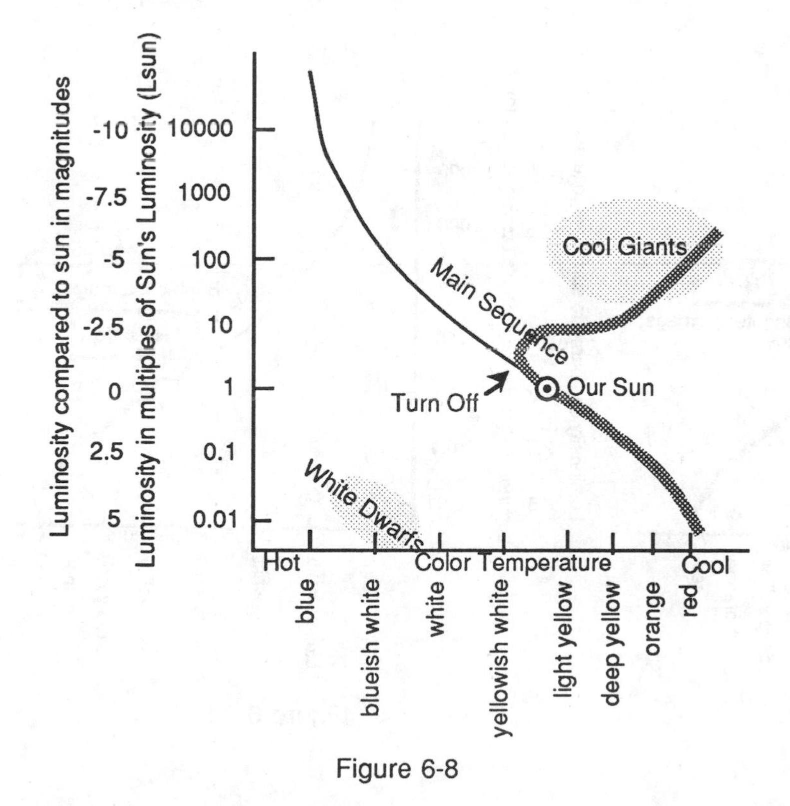

Figure 6-8

magnitudes of the brightest stars composing each cluster (Figure 6-9).

If the brightest stars are all blue, then you know them to lie on the upper main sequence and the cluster to be in its infancy (Figure 6-9a). If these brightest stars are a mix of red and blue stars, then the star cluster has reached its adolescence (Figure 6-9b). Middle aged star clusters reveal themselves with yellow/orange/cool giants dominating the brightest couple of magnitudes trailing off into whitish-yellow main-sequence stars (Figure 6-9c). In the oldest clusters, with the turn-off point having slid well down the main sequence, red/orange/yellow dominates the color of stars from the giants to the main sequence (Figure 6-9d).

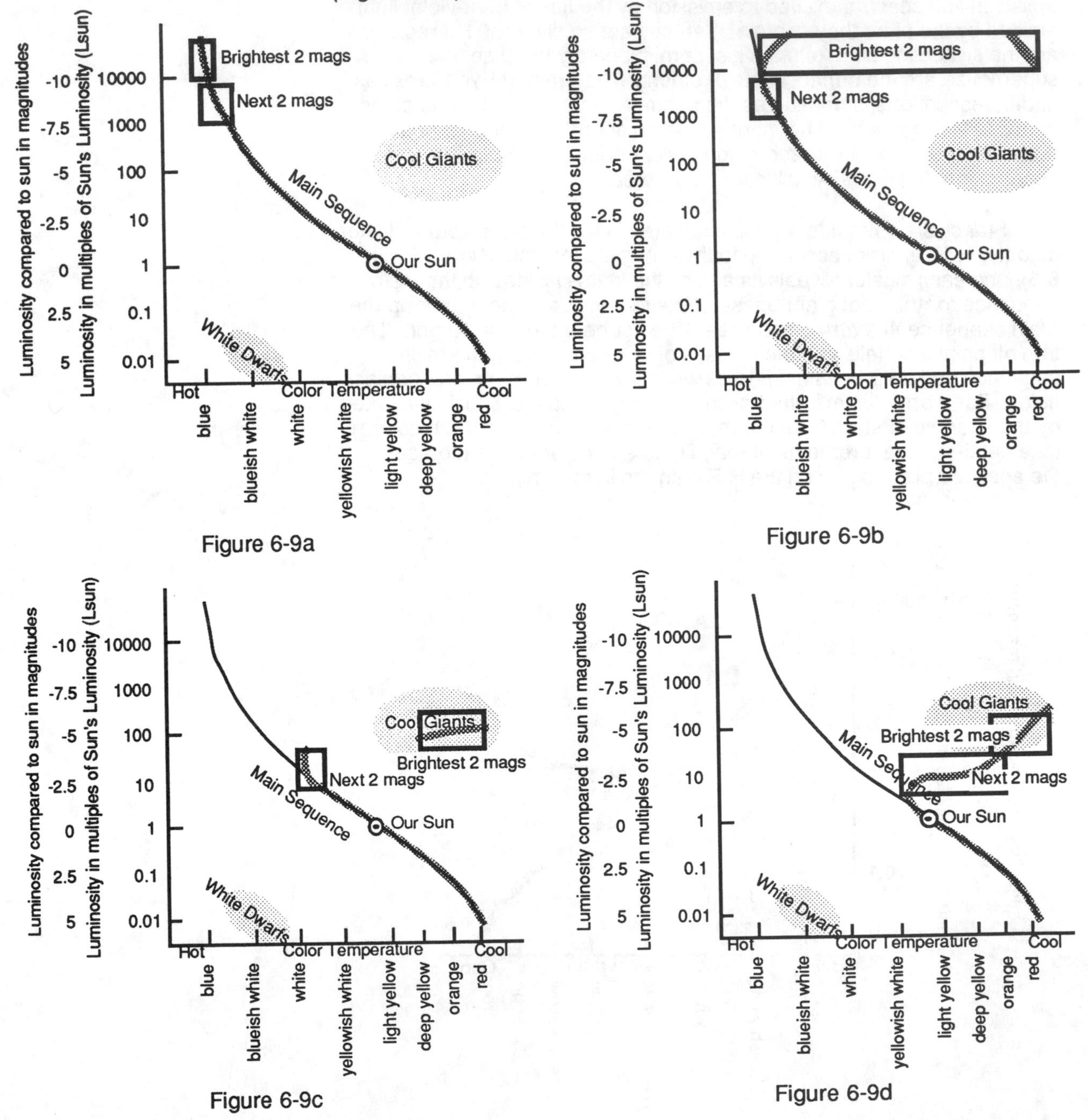

Figure 6-9a

Figure 6-9b

Figure 6-9c

Figure 6-9d

Observe a star cluster and divide its stars into the brightest two magnitudes and the next brightest two magnitudes, and all the rest. The colors of these groups reveal the star cluster's relative age (Table 6-2) and more.

Table 6-2: Colors and Ages of Star Clusters

Brightest Two Magnitudes	Next Two Magnitudes	Relative Age
Blue	Blue	Young
Blue & Red	Blue to White	Adolescent
Orange to Red	Yellowish White	Middle Aged
Orange to Red	Light Yellow	Old

By estimating a star cluster's age, you've estimated the turn-off point on its main sequence, the mass and luminosity that corresponds to the turn-off point, and thus the masses of the stars that are now leaving the main sequence and their fates. These revelations carry over into the nature of galaxies — the ultimate star clusters.

Observe a galaxy, and color reveals details of its evolution. The red glow of HII regions and the blue glare of young stars trace out the regions of star birth, while yellow-orange stars trace out the middle-aged and older components — both young and old revealed by the H-R diagram. From stars to galaxies, magnitude and color reveal their nature, their physics, their pasts, presents and futures.

Project 6-1
Seeing the H-R Diagram

Part A: Filtering Starlight

Although your color vision is senseless for faint stars, there is a simple technique you can use to perceive color with your night vision. Your night vision is sensitive to a range of wavelengths even though it produces an uncolored perception — pale white. You can recover the color information by artificially isolating the different wavelengths with colored filters. For example, a blue star and a yellow star of equivalent visual magnitude will look quite different when viewed through blue and yellow colored filters; the blue star will look brighter through the blue filter than it does through the yellow filter, while the yellow star will look brighter through the yellow filter than it does through the blue filter. Well, what about all the other color stars. There are more star colors than just blue and yellow. What filters would you use for red or orange or white stars?

It turns out that you can do a pretty good job with just two filters: blue and yellow. White stars appear to have equal brightness in both filters, orange stars glare in the yellow while only glimmering in the blue, whereas red stars barely show up in the blue. You can see the effects by using a blue and a yellow filter to judge the colors of the synthetic stars on the first slide.

These synthetic star images are produced by pin holes in cardboard backed with different colored filters. All the synthetic stars have been balanced to appear the same visual magnitude and all are too faint for your color vision. Compare each star's image through both the blue and the yellow filter. Remember to just compare the star to itself, not to other stars. Record in Table 1 whether the star is brighter in the blue or yellow filter, how much brighter and estimate the color of the star.

When you've finished estimating the colors of the stars with the blue and yellow filter, the intensity of the slide will be increased bit by bit until you can start to see color in the synthetic star images with your color vision. As color appears for the stars, note the real color and the order in which the colors became apparent in Table 1.

Questions

1) Rank star colors in order of ease to determine with the colored filters.

2) In which order did the true colors appear as the brightness of the star images increased? Describe.

Table 1

Star ID	Blue Filter	Yellow Filter	Estimated Color	Real Color	Order of Appearance

Part B: Star Clusters

The next several slides show color images of star clusters. For each cluster, count the number of stars; note the colors of the brightest and second brightest stars in Table 2; and sketch the locations of these stars in the schematic H-R diagrams provided. From the data in Table 2 and your sketches, estimate their relative ages and the point where stars are just evolving off the main sequence: the *main sequence turnoff.*

Questions

3) Which star cluster had the largest stars?

4) Which star cluster had the hottest stars?

5) List the star clusters from youngest to oldest. Is there any correlation between the main sequence turn-off point and the cluster's age?

6) In which star cluster are stars of the sun's mass just starting to leave the main sequence?

7) Compare the number of stars in each cluster to its age. Which star clusters had the most stars: youngest, middle aged, oldest? Why would such a correlation exist?

8) What are the relative distances of the star clusters?

Hint: Judging the angular size of a cluster will be tricky since they have no sharp edges; they just fade into the background stars. Still, using the law of perspective, the clusters with the largest angular size are closest, a cluster with half that angular size is twice as distant, etc. (assuming, of course, that all clusters have the same real size).

Table 2

Cluster ID	Total Number of Stars	Color of Brightest Stars	Color of Next Brightest Stars	Relative Age	Main Sequence Turn-off

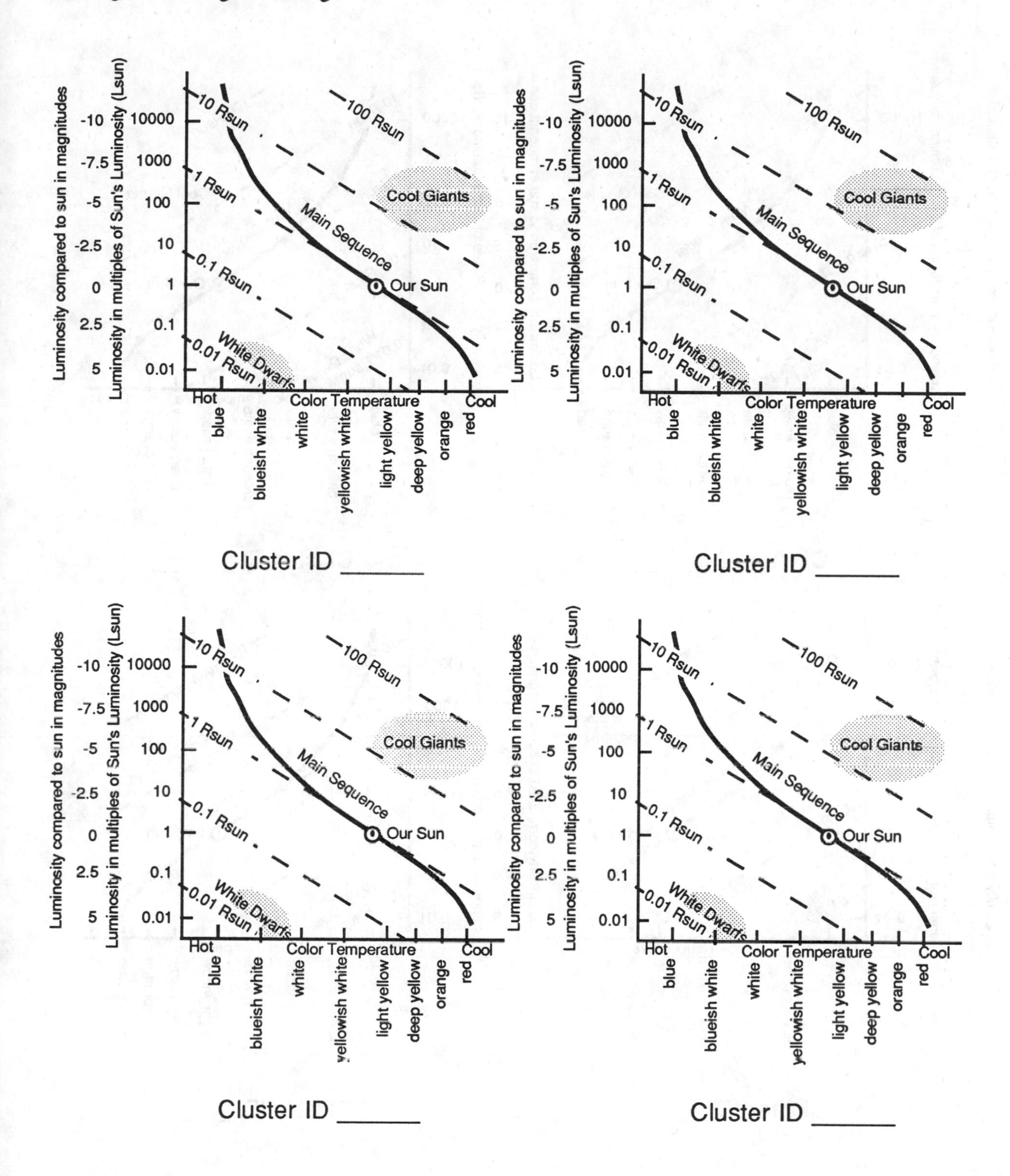
Luminosity compared to sun in magnitudes
Luminosity in multiples of Sun's Luminosity (Lsun)
-10
-7.5
-5
-2.5
0
2.5
5
10000
1000
100
10
1
0.1
0.01
10 Rsun
100 Rsun
1 Rsun
0.1 Rsun
0.01 Rsun
Cool Giants
Main Sequence
Our Sun
White Dwarfs
Hot
Color Temperature
Cool
blue
blueish white
white
yellowish white
light yellow
deep yellow
orange
red
Cluster ID ______
Luminosity compared to sun in magnitudes
Luminosity in multiples of Sun's Luminosity (Lsun)
-10
-7.5
-5
-2.5
0
2.5
5
10000
1000
100
10
1
0.1
0.01
10 Rsun
100 Rsun
1 Rsun
0.1 Rsun
0.01 Rsun
Cool Giants
Main Sequence
Our Sun
White Dwarfs
Hot
Color Temperature
Cool
blue
blueish white
white
yellowish white
light yellow
deep yellow
orange
red
Cluster ID ______
Luminosity compared to sun in magnitudes
Luminosity in multiples of Sun's Luminosity (Lsun)
-10
-7.5
-5
-2.5
0
2.5
5
10000
1000
100
10
1
0.1
0.01
10 Rsun
100 Rsun
1 Rsun
0.1 Rsun
0.01 Rsun
Cool Giants
Main Sequence
Our Sun
White Dwarfs
Hot
Color Temperature
Cool
blue
blueish white
white
yellowish white
light yellow
deep yellow
orange
red
Cluster ID ______
Luminosity compared to sun in magnitudes
Luminosity in multiples of Sun's Luminosity (Lsun)
-10
-7.5
-5
-2.5
0
2.5
5
10000
1000
100
10
1
0.1
0.01
10 Rsun
100 Rsun
1 Rsun
0.1 Rsun
0.01 Rsun
Cool Giants
Main Sequence
Our Sun
White Dwarfs
Hot
Color Temperature
Cool
blue
blueish white
white
yellowish white
light yellow
deep yellow
orange
red
Cluster ID ______

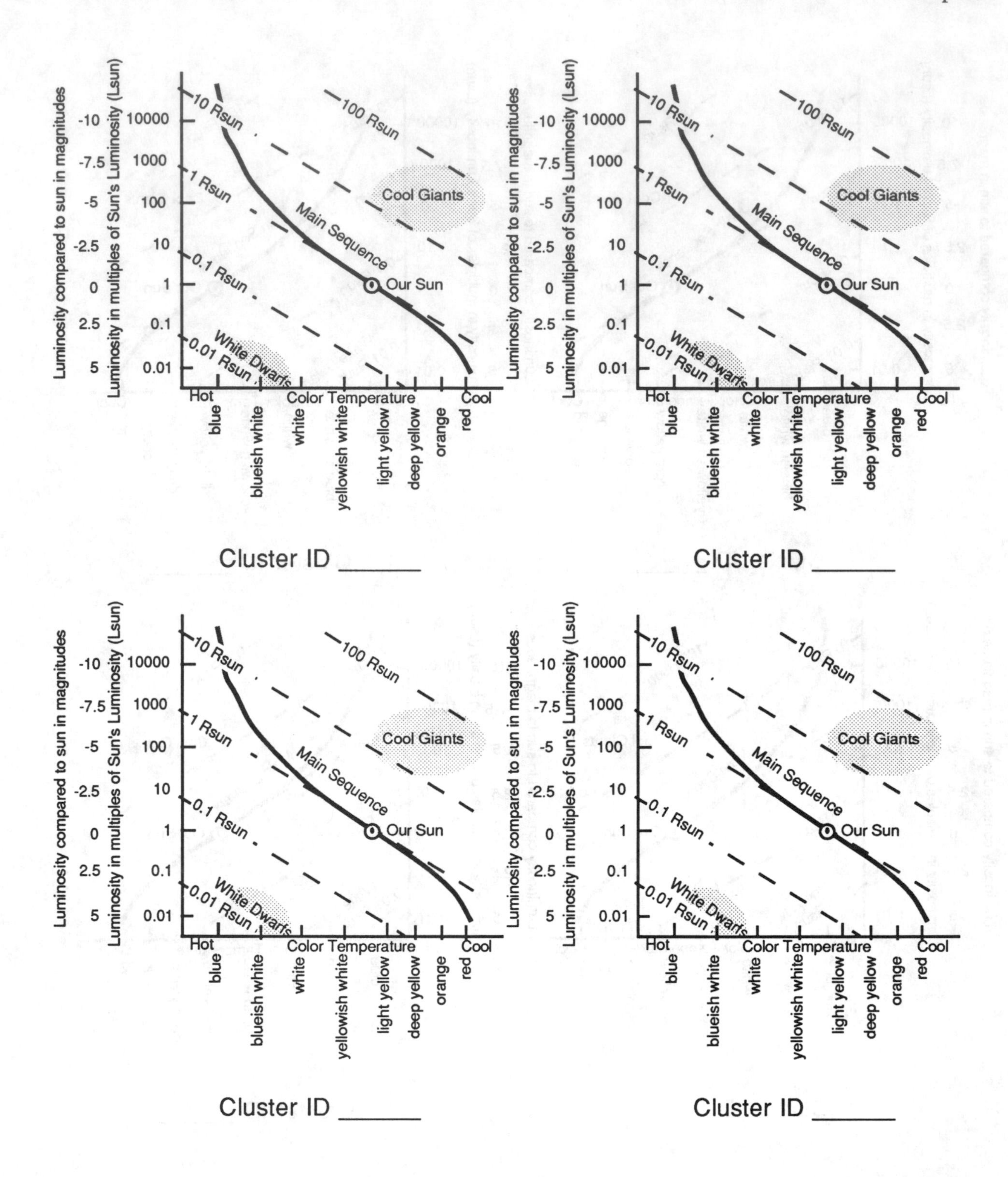

Luminosity compared to sun in magnitudes
Luminosity in multiples of Sun's Luminosity (Lsun)
-10
-7.5
-5
-2.5
0
2.5
5
10000
1000
100
10
1
0.1
0.01
10 Rsun
100 Rsun
1 Rsun
0.1 Rsun
0.01 Rsun
Cool Giants
Main Sequence
Our Sun
White Dwarfs
Hot
Color Temperature
Cool
blue
blueish white
white
yellowish white
light yellow
deep yellow
orange
red
Cluster ID ______
Cluster ID ______
Cluster ID ______
Cluster ID ______

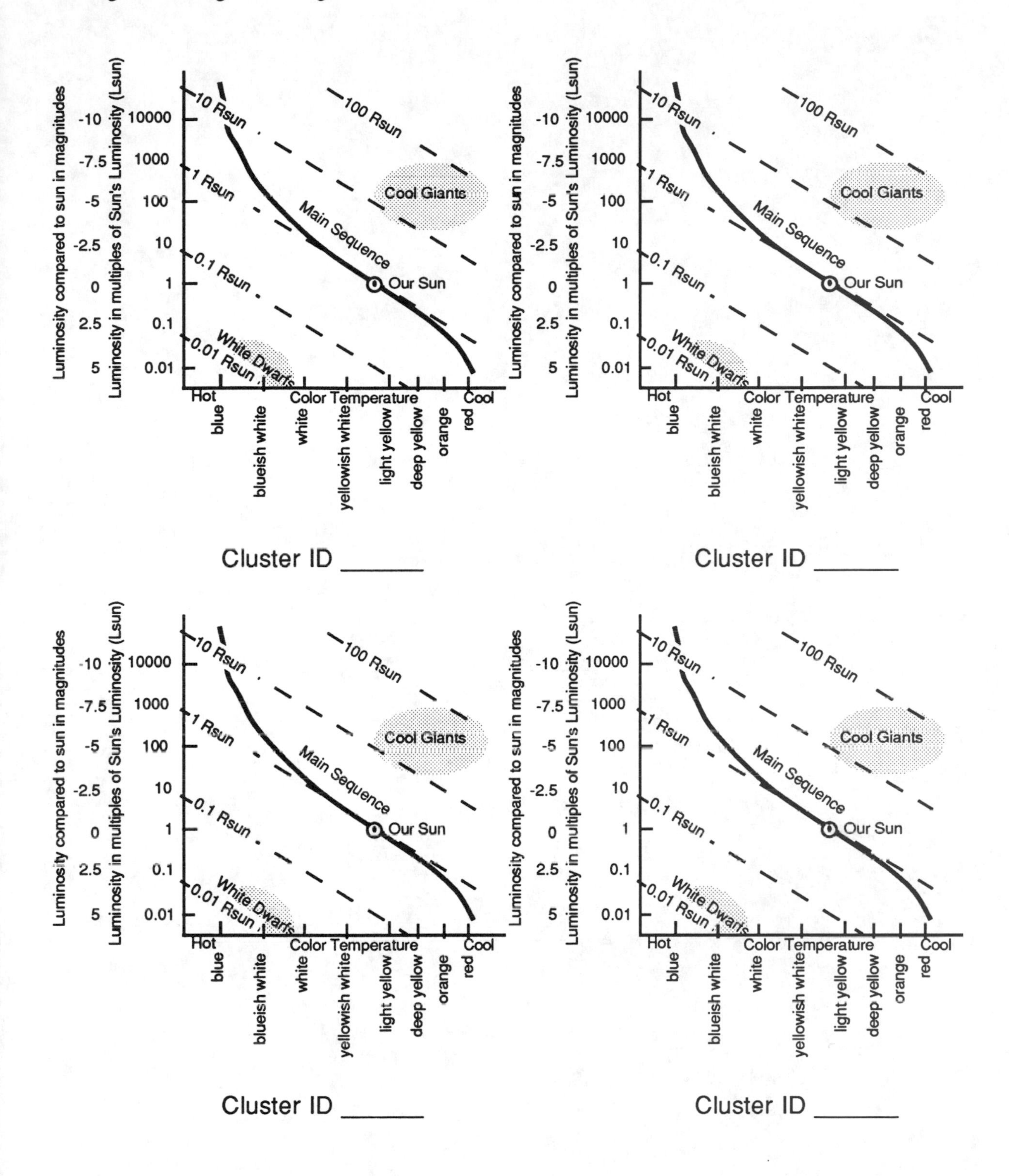
Luminosity compared to sun in magnitudes
Luminosity in multiples of Sun's Luminosity (Lsun)
-10
-7.5
-5
-2.5
0
2.5
5
10000
1000
100
10
1
0.1
0.01
10 Rsun
100 Rsun
1 Rsun
0.1 Rsun
0.01 Rsun
Cool Giants
Main Sequence
Our Sun
White Dwarfs
Hot
Color Temperature
Cool
blue
blueish white
white
yellowish white
light yellow
deep yellow
orange
red
Cluster ID ______

Part C - Galaxies

Now its time to turn your attention to the most spectacular collections of stars (and other things), galaxies. On the next several slides, you will see galaxies of several types. For each galaxy:

1) Check off what it contains (that you can see): blue stars, yellow stars, dust, open star clusters, globular star clusters, HII regions, in Table 3.

   Hint: Remember that most galaxies are so distant that you won't see stars individually; you'll just see their combined glow. Dust will show itself by blocking starlight, HII regions, big enough to be seen individually, by their warm red glow, star clusters as just barely resolved patches of starlight.

2) Examine the distribution of blue, yellow and orange stars and record on a sketch of each galaxy. On the sketch also note the location of dust and HII regions.

3) Count the number of HII regions, open star clusters, globular star clusters each galaxy contains and record that result in Table 3.

Questions

8) Which types of galaxies had blue stars? Which types of galaxies had yellow-orange stars? Which types of galaxies had both colors of stars?

9) Which types of galaxies contained dust? Which types of galaxies contained HII regions? Is there any correlation between the two? Explain.

10) Which colors of stars were associated with dust in the galaxies that contained dust? Is this correlation reasonable? Explain.

11) Which colors of stars were most closely associated with the HII regions in galaxies that contained HII regions? Is this correlation reasonable? Explain.

12) Was there any correlation between the number of HII regions and the galaxy type?

    Hint: Look for relationships within a galaxy type (e.g., Sa to Sd) and across galaxy types (E vs S vs Irr).

13) Which type of galaxy contained the greatest number of globular clusters? What colors of stars did this type of galaxy contain? What else did this type of galaxy contain? Are these correlations reasonable? Explain.

Table 3

Galaxy Type	Blue Stars	Yellow-Orange Stars	Dust	HII Regions	Open Star Clusters	Globular Star Clusters

Project 6-2
Star Search

First set up and align your telescope using the coordinated polar alignment technique to get an accurate polar alignment. The four main steps are as follows:

Step 0: Quick and Dirty Polar Alignment
Step 1: Find Polaris by its Coordinates
Step 2: Align Mounting to Center Polaris in Telescope
Step 3: Check Right Ascension Dial
Step 4: Repeat until Perfect

Now that you are set up and aligned, start finding and observing the multiple stars assigned by your instructor from Table 1 from their right ascension and declinations. Your observation for each star should include the information described in Box 6-1.

Observe, as outlined above, selected stars from the list. If you have trouble seeing or resolving certain stars, try and try again. Describe any problems you had in finding them and how you overcame those problems.

Tricks of the Trade

1) *The trick to determining NESW as seen in the eyepiece is to let the star's motion tell you which way is west.*

   A) *Center the star in the eyepiece.*

   B) *Turn off the clock drive*

   C) *The stars will drift to the west side of the eyepiece.*

   D) *Put your finger on the side of the eyepiece the stars move ttowards. That is the west side of the eyepiece.*

   E) *Label that side of your sketch using W for the west.*

   F) *Restart your clockdrive so that you don't get too far behind and check the right ascension.*

   G) *On your sketch, east is directly opposite west. Determining north or south depends on whether or not you are using a diagonal eyepiece holder.*

If you are not using a diagonal, then south is 90° counter-clockwise from east, and north is 90° counter-clockwise from west.

If you are using a diagonal, then south is 90° clockwise from east, and north is 90° clockwise from west.

2) *Always use a low-power eyepiece to acquire and center the star in the telescope. In your writeup, explain the advantage of using low power for getting the star into your eyepiece.*

3) *Develop a sense for how bright stars of different magnitude look through different eyepieces and the finder.*

4) *Develop a sense of how different separations for multiple stars look through different eyepieces and the finder.*

5) *To see color in bright stars, defocus the images slightly.*

6) *Check the Multiple Star System pages of The Trained Eye Star Atlas for additional information on each multiple star.*

7) *The rule of thumb for selecting an eyepiece is to use an eyepiece with a focal length equal to or less than the separation of the components of a multiple star. That is, if the separation is 20", then a 20-mm eyepiece will just let you see both, a 10-mm eyepiece will let you see them easily. This eyepiece selection rule assumes novice eyes, an f-ratio of 6 to 8 and objective diameter of 6 to 10 inches. It also assumes that both components are of equal magnitude. If they are 5 magnitudes or more different, you'll need even more magnification.*

8) *Always check your circles before leaving a star. If the right ascension dial is wrong, correct it. If the declination dial is wrong, calculate the Dec Error.*

9) *Help yourself by writing down problems in your notes as you encounter them.*

Questions

1) What environmental factors do you think might affect visibility (e.g., altitude, seeing, transparency, etc.) ? Explain the effect of each.

   Hint:: Give examples from your observations.

2) Was there a systematic difference between the temperatures of the brightest component of a multiple star and the temperatures of the faintest component stars? Describe.

   Hint: The colors of stars range from blue hot to red cool:

blue blue-white white yellowish-white light -yellow deep-yellow orange red

Hint: List the number of stars of each color for the brightest component and for the faintest components of a multiple star system. Color categories are: blue, blue-white, yellowish-white, light yellow, yellow, orange, red, no color. For faint stars white means no color.

3) How many stars showed color?

Hint: Remember that for faint stars white means NO color so don't count faint white stars.

4) What was the most common star color?

Hint: See Hint for question 2.

5) What was the magnitude of the faintest component star of a multiple star in which you saw color? What was that color?

6) Which multiple star had the component star with the coolest temperature? Was it the brighter or fainter component star?

7) Use the colors and magnitude differences of the multiple stars to place their components on the H-R diagram. For those where only the brighter component star shows color, use the H-R diagram to deduce the color of the fainter component star.

Hint: When both stars show color:

1) *Draw a vertical line on the H-R diagram at the color of each (each star's color line).*
2) *PLace the brighter star on its color line at the top the H-R diagram .*
3) *Place the fainter on its color line below the brighter star by the amount of the magnitude difference between the two stars.*
4) *Slide the two stars down their color lines, maintaining their vertical seperation, until they align with features of the H-R diagram (i.e. cool giant, main sequence, white dwarf).*

When only the brighter star shows color:

1) *Draw a vertical line on the H-R diagram at the color of the brighter component star (its color line).*
2) *Place the brighter star on its color line at the top of the H-R diagram (the bead).*
3) *Place a horizontal line straight across the H-R diagram below the brighter star by the amount of the magnitude difference between the two component stars.*
3) *Slide the two stars (bead and horizontal line) down the H-R diagram until the brighter star aligns with a feature of the H-R diagram (i.e. cool giant or main sequence) .*

5) *Where the horizontal line (representing the fainter component star) crosses a feature of the H-R diagram (I.e. Main sequence, white dwarf) identifies its color.*

Label the stars on the sketch with the star name followed by the letter A, B, C or D from brightest component to faintest. For example, the two components of γ Del would be labeled γ Del A and γ Del B.

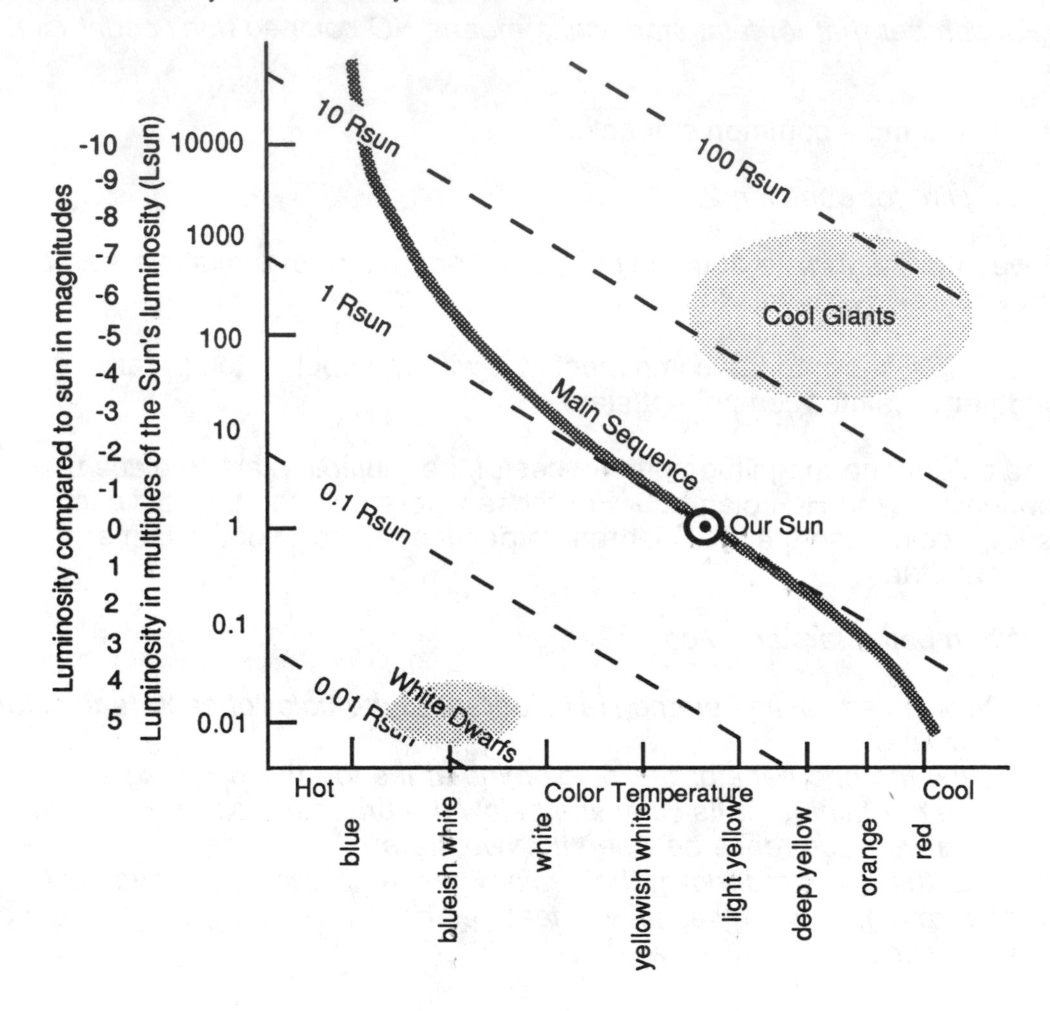

8) List the number of white dwarfs, main sequence and giant stars.

9) Which multiple star(s) had the biggest star?

10) Which multiple star(s) had the youngest star?

11) Which multiple star(s) had the most evolved star?

Table 1: Multiple Star Systems

Bayer Flamsteed Name	RA h	RA m	Dec °	Dec '	Visual Magnitudes		Separation "
γ And	2	04	42	20	2.2	5.1	9.8
α UMi	2	20	89	16	2.0	8.9	18.4
α Cet	3	02	4	05	2.5	5.6	960.0
ζ Per	3	54	31	53	2.9	9.4	12.9
ε Per	3	58	40	00	3.0	8.1	9.0
β Ori	5	14	-8	11	0.2	7.0	9.2
δ Ori	5	32	-0	22	2.5	6.9	52.8
θ Aur	6	00	37	13	2.6	7.1	3.4
ε CMa	6	59	-28	58	1.6	8.1	7.4
α Gem	7	35	31	54	2.0	2.9	2.2
α Leo	10	08	11	58	1.3	7.6	176.5
γ Leo	10	20	19	52	2.2	3.5	4.3
δ Crv	12	30	-16	30	3.0	8.4	24.2
α CnV	12	56	38	18	2.9	5.4	19.7
ζ Uma	13	24	54	56	2.3	3.9	14.4
ε Boo	14	45	27	04	2.7	5.1	2.9
α Lib	14	51	-16	02	2.9	5.3	231.0
β Sco	16	5	-19	48	2.7	4.9	13.7
η Dra	16	24	61	31	2.9	8.8	6.1
α Sco	16	30	-26	25	1.2	6.5	2.9
α Her	17	15	14	23	2.5	5.0	4.6
β Cyg	19	31	27	57	3.1	5.1	34.8
β Cap	20	21	-14	46	3.1	6.2	205.0
β Cep	21	29	70	33	3.2	7.8	13.6

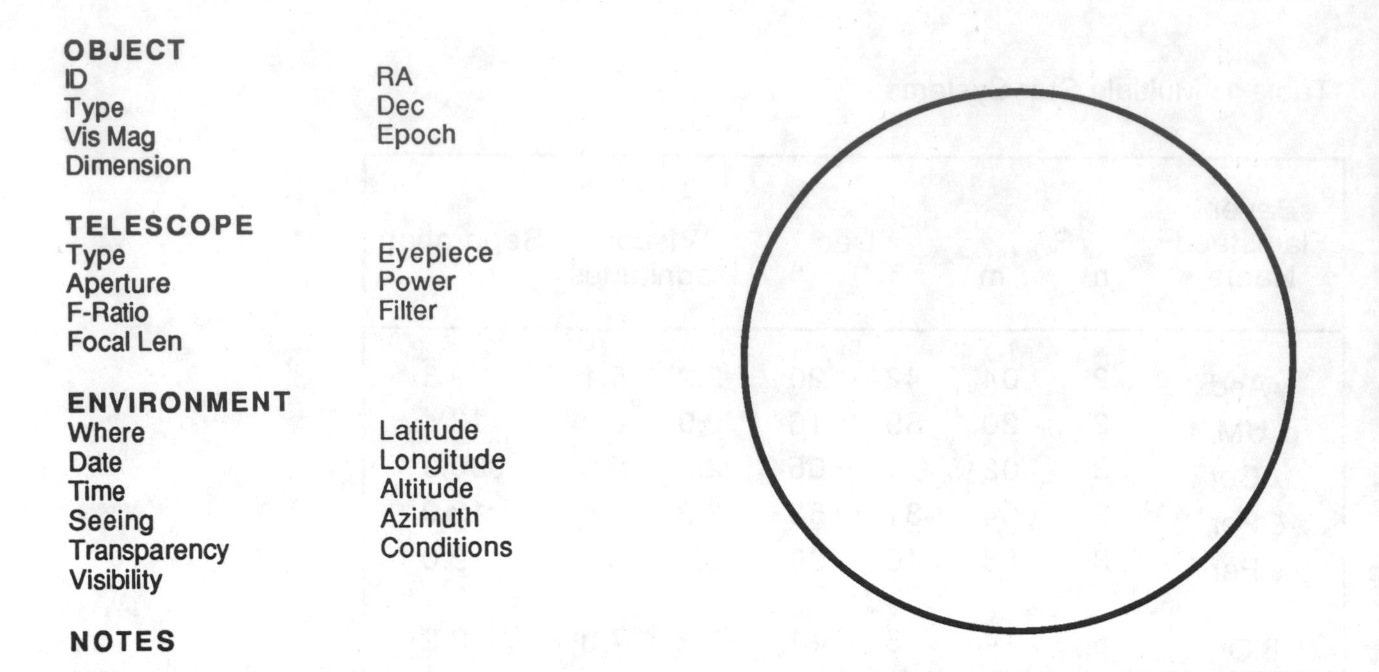

OBJECT

ID	RA
Type	Dec
Vis Mag	Epoch
Dimension	

TELESCOPE

Type	Eyepiece
Aperture	Power
F-Ratio	Filter
Focal Len	

ENVIRONMENT

Where	Latitude
Date	Longitude
Time	Altitude
Seeing	Azimuth
Transparency	Conditions
Visibility	

NOTES

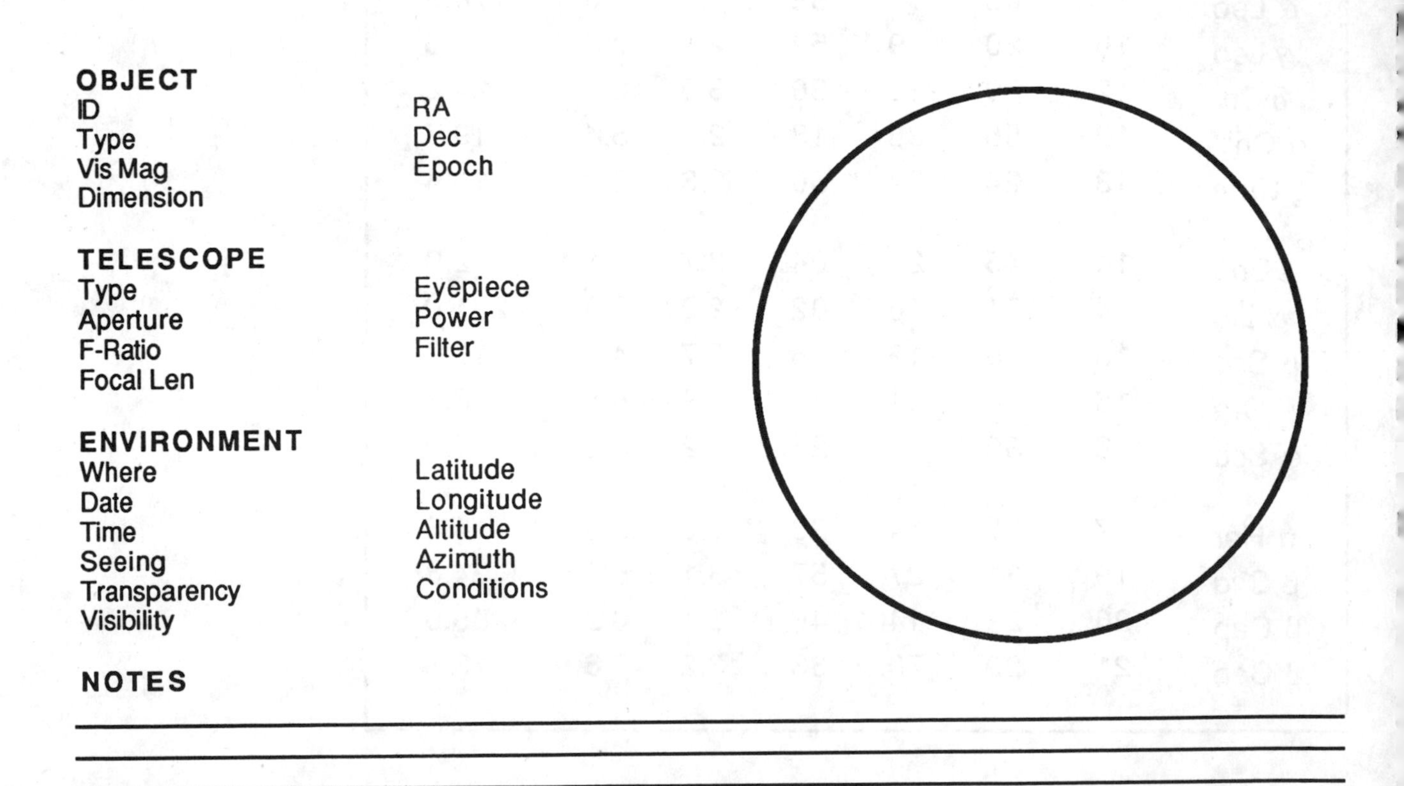

OBJECT

ID	RA
Type	Dec
Vis Mag	Epoch
Dimension	

TELESCOPE

Type	Eyepiece
Aperture	Power
F-Ratio	Filter
Focal Len	

ENVIRONMENT

Where	Latitude
Date	Longitude
Time	Altitude
Seeing	Azimuth
Transparency	Conditions
Visibility	

NOTES

OBJECT

ID	RA
Type	Dec
Vis Mag	Epoch
Dimension	

TELESCOPE

Type	Eyepiece
Aperture	Power
F-Ratio	Filter
Focal Len	

ENVIRONMENT

Where	Latitude
Date	Longitude
Time	Altitude
Seeing	Azimuth
Transparency	Conditions
Visibility	

NOTES

OBJECT

ID	RA
Type	Dec
Vis Mag	Epoch
Dimension	

TELESCOPE

Type	Eyepiece
Aperture	Power
F-Ratio	Filter
Focal Len	

ENVIRONMENT

Where	Latitude
Date	Longitude
Time	Altitude
Seeing	Azimuth
Transparency	Conditions
Visibility	

NOTES

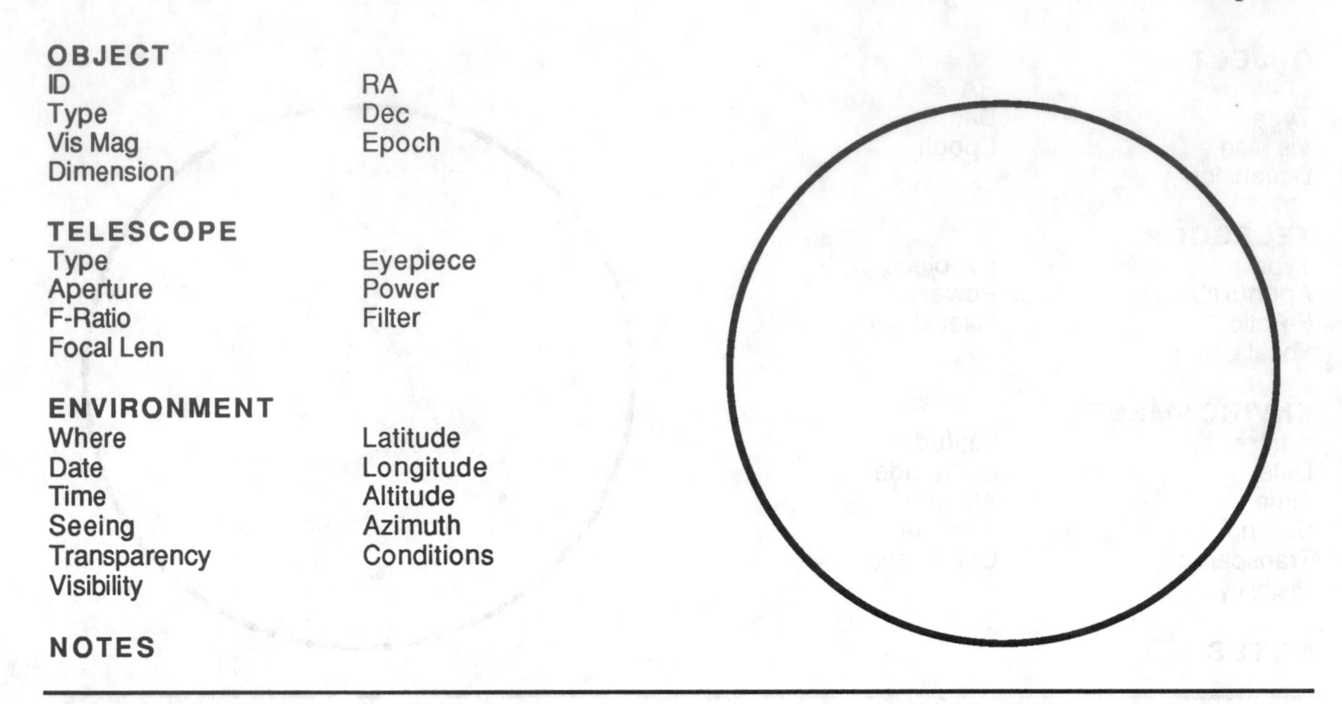

OBJECT

ID	RA
Type	Dec
Vis Mag	Epoch
Dimension	

TELESCOPE

Type	Eyepiece
Aperture	Power
F-Ratio	Filter
Focal Len	

ENVIRONMENT

Where	Latitude
Date	Longitude
Time	Altitude
Seeing	Azimuth
Transparency	Conditions
Visibility	

NOTES

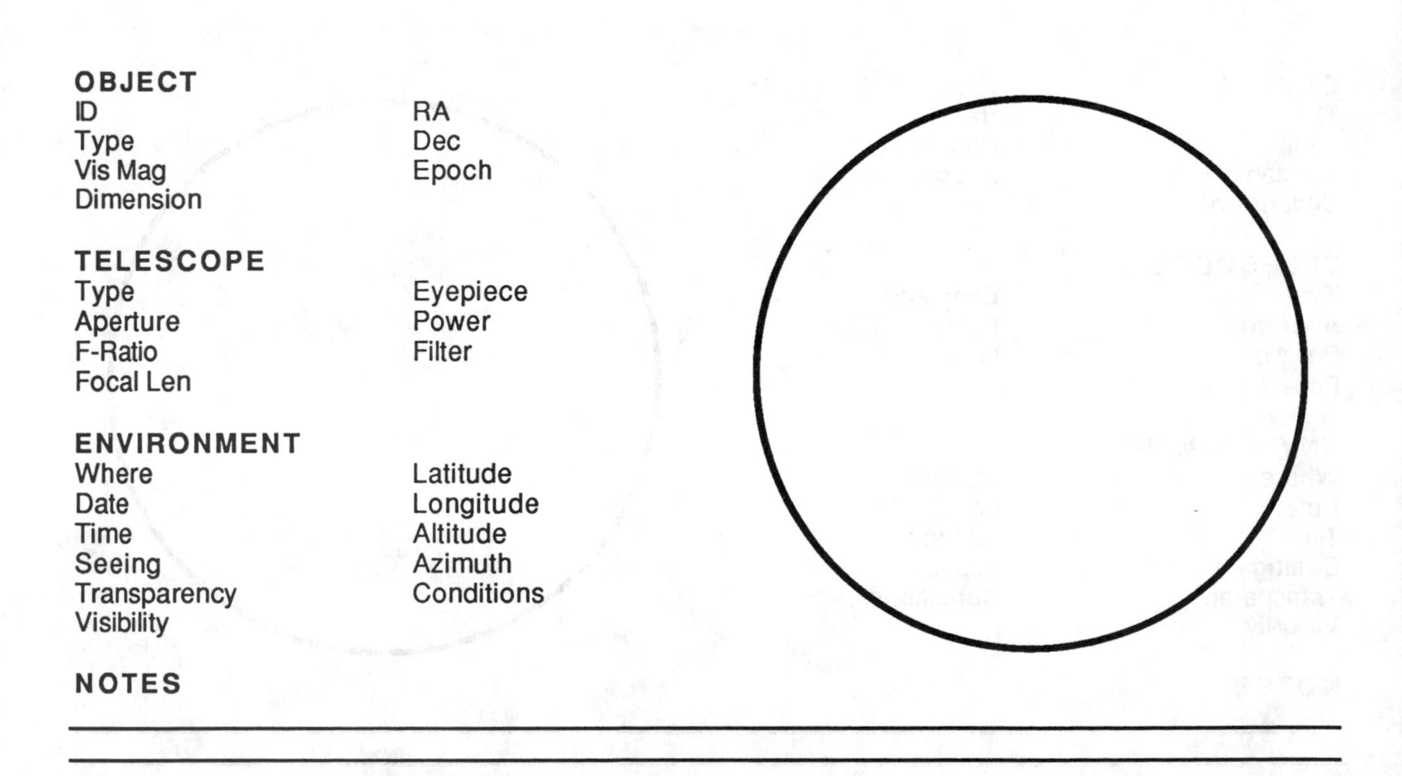

OBJECT

ID	RA
Type	Dec
Vis Mag	Epoch
Dimension	

TELESCOPE

Type	Eyepiece
Aperture	Power
F-Ratio	Filter
Focal Len	

ENVIRONMENT

Where	Latitude
Date	Longitude
Time	Altitude
Seeing	Azimuth
Transparency	Conditions
Visibility	

NOTES

OBJECT

ID | RA
Type | Dec
Vis Mag | Epoch
Dimension

TELESCOPE

Type | Eyepiece
Aperture | Power
F-Ratio | Filter
Focal Len

ENVIRONMENT

Where | Latitude
Date | Longitude
Time | Altitude
Seeing | Azimuth
Transparency | Conditions
Visibility

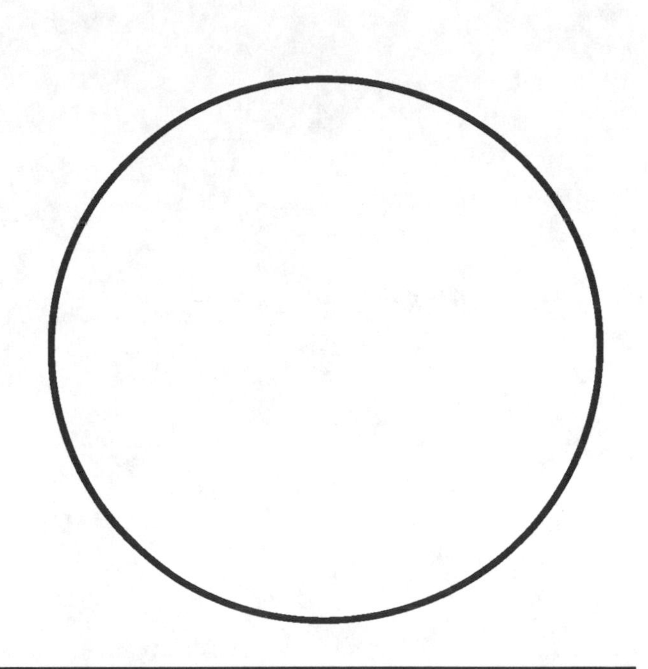

NOTES

OBJECT

ID | RA
Type | Dec
Vis Mag | Epoch
Dimension

TELESCOPE

Type | Eyepiece
Aperture | Power
F-Ratio | Filter
Focal Len

ENVIRONMENT

Where | Latitude
Date | Longitude
Time | Altitude
Seeing | Azimuth
Transparency | Conditions
Visibility

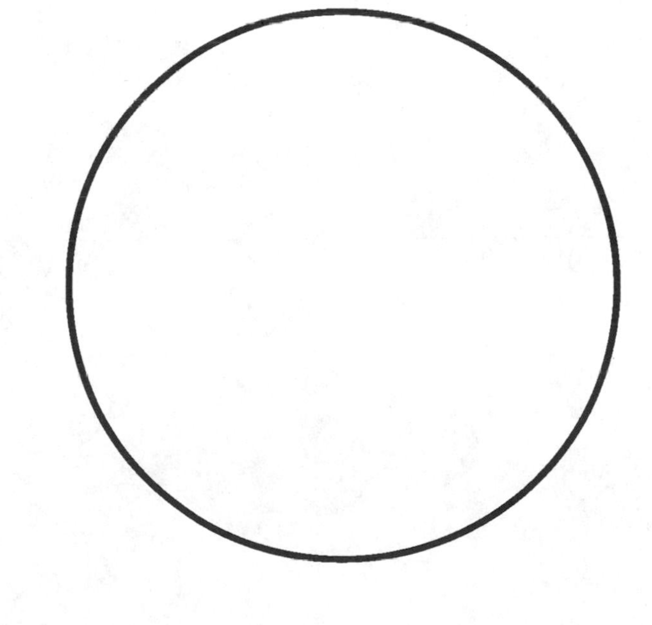

NOTES

Chapter 7
Observing

You trained in more skills than ust the properties of your telescope and setting up an equatorial mounting in Chapter 4. You learned to recognize different magnitude stars through different size telescopes and eyepieces, how to compare what you see on a star chart to what you see through an eyepiece, how to move your telescope one direction and think the other, how to move north/south/east/west instead of up/down/left/right — *you began to observe.*

In Chapter 5 and 6 you learned how the message in starlight, revealed in its spectral splash across the canvas of the Hertzsprung-Russell diagram, proves that stars are indeed faraway suns — in vindication of Giordano Bruno. You see revealed upon the Hertzsprung-Russell diagram the measures of stars — their sizes, temperatures, luminosities, and fates. Individually and in their natural groupings as multiple stars, star clusters, galaxies, and clusters of galaxies, your observation of color and magnitude reveals the universality of physics postulated by Newton. You've seen the beauty of physics in the observation.

This final chapter completes your apprenticeship in observation by presenting a final compilation of observing tricks of the trade and challenging you to apply all your skills and knowledge as you wonder among the stars.

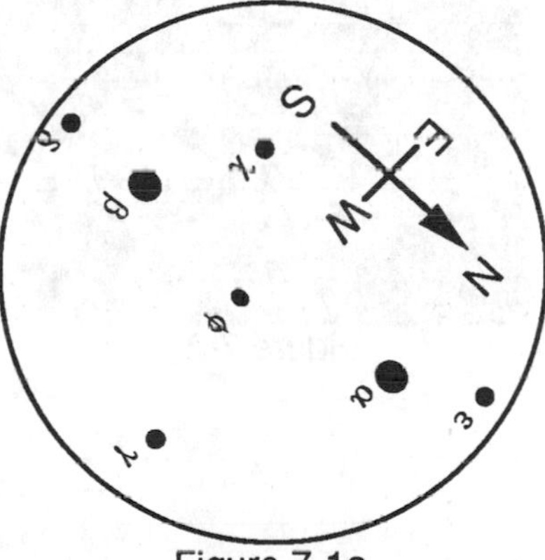

Figure 7-1a

Seeing the Big Picture

Remember the finder telescope? That little wide-field telescope strapped to the side of your main telescope that lets you see some 4 to 5° field of the sky and stars down to 8th to 9th magnitude? With the finder you see a large enough patch of the celestial sphere to compare to what you see to the star maps in The Trained Eye Star Atlas. If the star map shows a pattern of three or four stars near what you're looking for, then you should see the same pattern of three or four stars in the finder. How do you make the comparison? Hold the star map up next to the finder. Of course this requires you to have the star map correctly oriented to what you see in the finder. For that, you use the same technique you do for the main telescope to determine north, east, south, and west in the finder's eyepiece (Project 4-2), and then hold The Trained Eye Star Atlas (Figure 7-1a) to match the stars in the finder (Figure 7-1b).

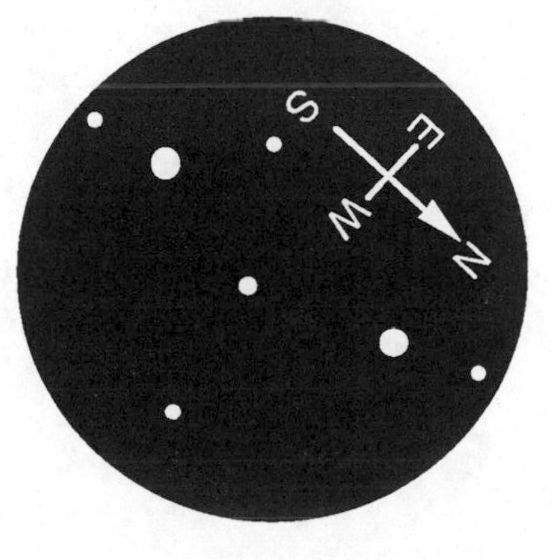

Figure 7-1b

In addition, you can extend the field of view of your main scope by *sweeping*. Memorize what you see in the eyepiece, keep watching, and slew the telescope just off of what you memorized. If you don't see what you're looking for, slew the opposite way back to where you started. Repeat this process for the other slewable directions until you see what you're after.

If you don't see what you're after, you can use another version of slewing called *spiral box slewing*. You start the same way, memorizing your starting point, slew off a little in one direction while watching and stop. Instead of slewing back to your starting point, slew crosswise a little and stop. Then slew back past your starting point and stop, slew crosswise the other direction and stop. Continue spiraling out from your starting point (Figure 7-2) until you find your object. If you don't find it after a while, then you probably have a major error that requires major fixing as described in Chapter 4.

The key to slewing is memory. Remembering which directions you've slewed, how far, what you see in the eyepiece each time you stop. What, however, do you do when you still end up with three or four stars in the eyepiece that might be the one you're looking for? Check them all and see which one best matches the description of the star you're looking for.

Figure 7-2

Hop to It

You learned in Chapter 4 to keep track of your declination and right ascension dials and to check them repeatedly and the effects of a bad polar alignment or a bent polar axis. Even if they're right and you still don't see anything, how do you show you're in the right place? The solution: map out the route starting from something that's easy to find to the object you're trying to find. Following this route, hopping from star to star provides visual feedback that you're on the right track. This technique is called *star hopping* and goes like this:

1) Know What You're Looking for:

   It always helps to know what you're looking for before you go looking. So before going hunting, you should read the appropriate notes in The Trained Eye Star Atlas that describe what you're trying to find. If it is a NGC object, you can find a description of what it is (open cluster, globular cluster, planetary nebula, HII (pronounced H two) region, galaxy, supernova remnant, etc.) and what it should look like. If it's a multiple star system, you'll find information on the apparent visual magnitudes of its component stars, their separations, colors, etc.

2) Pick a Bright Star as Your Starting Point.

   Start from something that's easy to find — a bright star (0th to 2nd magnitude). Pick a bright star that is reasonably close to what you're trying to find, within 30° or so (don't pick one way on the other side of the sky). There may be several to choose from, each of which is the starting point for a different route to the faint object.

3) Pick Way Stations From the Bright Star to the Object.

Pick way stations along the path from each bright star to the faint object. These should be middling brightness (2nd to 4th magnitude) stars. Don't pick way stations that are fainter than what you're trying to find. The distance between way stations should be kept small, never more than about 10° (remember this corresponds to 40 minutes of right ascension at the celestial equator, and progressively greater amounts of right ascension as you approach the north or south celestial poles). You should never need more than two or three way stations.

4) Use Easy Slews.

At this point you will have several possible paths to the faint object. Now you have to determine which of these paths is best. The best path is that which has the *easiest slews* .

Easy slews are rotations of the telescope around one axis at a time (e.g., a north/south slew in declination is easy, an east/west slew in right ascension is easy). Conversely, a slew that requires you to move the telescope great distances in both declination and right ascension at the same time is a *hard slew* as you'll no doubt find out.

Sometimes you can't find any single easy slew from one way station to another. In this case you might have to use *staircase slews*, that is a slew in declination followed by a slew in right ascension (e.g. slew north/south in declination by some amount, then east/west in right ascension by some amount). It is still best to keep one of the two slews as close to an easy slew as possible since the further you slew, the greater the potential for error. Always plan to slew in declination first if you can since you can lock the declination axis after you've slewed.

5) Prove You're There

So you get to where you think the object is, but you either don't see it, or you see something that you're not sure is it. How do you prove that you're at the right place? Simply by looking around for other stars that should, according to the star atlas, be near it. Always pick *proof* stars that are brighter than what you're trying to find. Sometimes these proof stars will all be visible in the finder at the same time and its a simple matter of comparing what you see through the finder to what you see in the atlas.

Most times, however, you will not find any recognizable pattern of stars around the object. In this case, find some easily recognizable object or pattern of stars that is an easy slew away, and plan on using it as proof. You should be able to predict from The Trained Eye Star Atlas how far and in which direction you have to slew from the faint object to get to your proof object.

One proof object is not sufficient unless it is so unique looking that there can be no mistake (remember there are lots and lots of faint stars out there). You need at least two proof objects for positive confirmation.

This process of star hopping can also be adapted to an alt-azimuth mounting. The only catch is that the motions of an alt-azimuth mounting are in human coordinates, not celestial coordinates — you move the telescope up, down, left, and right not north, south, east, and west. The basic principle remains the same; start with an easy to find star, use way stations to help you along, directions and distances on the star map determine directions and distances to move your telescope. However, you have to transform The Trained Eye Star Atlas into human coordinates by holding it so it matches the constellations and then determining the up/down/left/right slews required to get from your starting point to the object of interest (Figure 7-3).

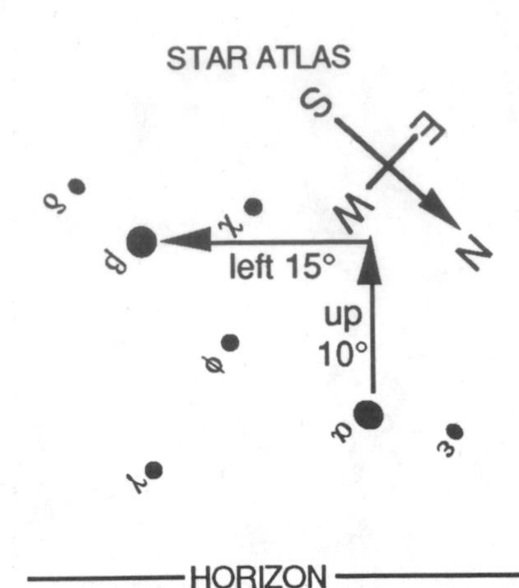

Figure 7-3a

SKY

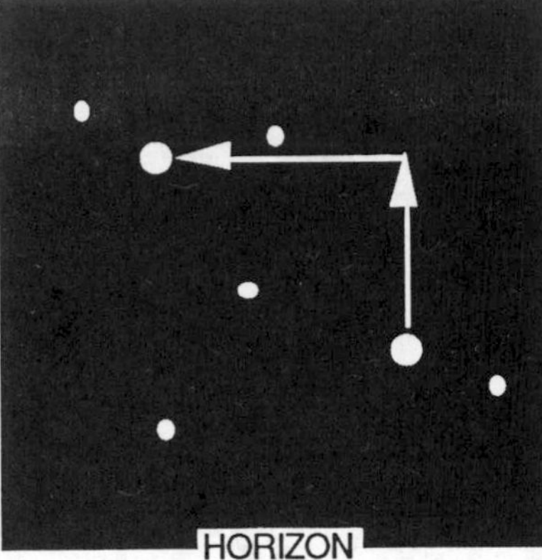

Figure 7-3b

Ignorance Isn't Bliss

The more you know about what you're looking for, the easier it is to find. The Trained Eye Star Atlas divides celestial objects into different types and contains additional information for each beyond right ascension and declination: *apparent visual magnitude, dimensions*, and *description*. How you use this information depends upon the type of celestial object, as will be discussed in the next section.

The Visibility of Celestial Objects

The *visibility* of a non-stellar object, whether you can see it or not, depends upon its apparent visual magnitude (obvious) and upon its dimensions (not so obvious). A bright object spread all across the sky (the Milky Way) is much less visible than a faint object concentrated into a point (a star). You need a measure of faintness independent of size to determine the visiblity of non-stellar objects. Such a measure is called surface brightness. Calculate the approximate area of the non-stellar object from its dimensions in The Trained Eye Star Atlas by:

$$\text{Area} = \text{length} * \text{width} \qquad \text{for oblong objects or}$$

$$\text{Area} = \pi * \text{radius}^2 \qquad \text{for round objects}$$

Divide area into apparent visual magnitude and you have surface brightness (in magnitude per square arc-minute).

$$\text{Surface Brightness} = \frac{\text{Apparent Visual Magnitude}}{\text{Area}}$$

A good example of surface brightness are clouds. Earthbound clouds and interstellar *nebulae* (Latin for clouds) glow all over, not just at one point. If two nebulae have the same apparent visual magnitude, then the smaller will be easier to see (i.e., more visible). If two nebulae have the same surface brightness, both will be equally visible regardless of their differing sizes. To determine your limiting visibility for non-stellar objects, just find the object with the faintest surface brightness you can see for your sky, telescope, and eyes.

Fading into Twilight

Sky glow, besides supplying another example of surface brightness, has an especially deleterious effect on the visibility of non-stellar objects (be they nebulae or star clusters). Figure 7-4 shows two photographs of the Pleiades open star cluster. Both photographs were taken with the same telescope and the same exposure time. Guess which one was taken in Los Angeles? Guess which one matches the description in The Trained Eye Star Atlas? The brighter the sky glow, the less you should expect what you see to match its description.

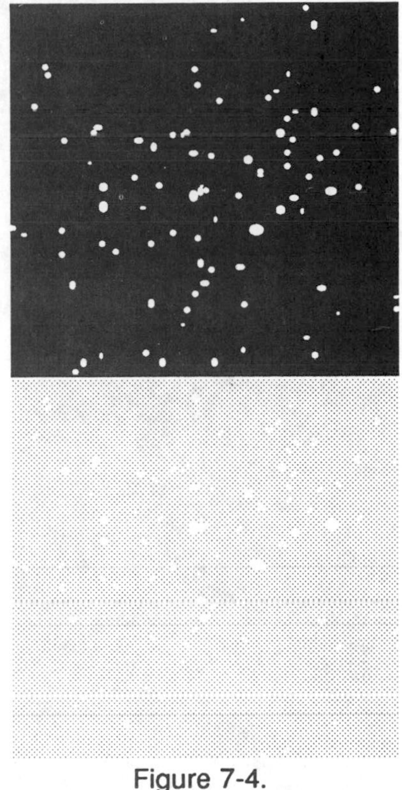

Figure 7-4.

Magnification can help reduce the effects of a sky glow for stars whether singly or in clusters, as you found in Project 4-1. As you increase magnification, the sky glow gets darker while the stars stay the same brightness. As a result, stars become more visible. This trick works because each star's light is concentrated into a point, while sky glow is spread over an area.

As you go to higher power, each detector cell in your eye's retina sees less area of the sky. Since the surface brightness of the sky is per square arc-minute, if each detector cell sees fewer square arc-minutes, then each detector cell sees less sky glow and the sky appears to darken. Fortunately, stars are point sources, so their light doesn't spread out over more detector cells as you increase magnification. Each cell sees the same light as always but against a darkening background; the *contrast* between the stars and the sky improves (Figure 7-5).

All this is true to a point. Use enough magnification and stars cease to look like points — they start looking like blur disks (Chapter 3). When you see their blur disks, you've reached the end of the game, for increasing magnification spreads the star light out across detector cells just as fast as it spreads out the sky glow.

Magnification can also help for interstellar nebulae (here nebulae is used in the generic non-stellar object sense of planetary nebulae, supernovae, H II regions, and galaxies — all objects that don't appear starlike). As you increase magnification, nebulae look bigger but not brighter. You are spreading the same amount of brightness over a larger surface of your retina just as fast as you dilute the sky glow over more detector cells. As a result, the contrast between the nebulae and the sky glow remains constant. Fortunately, the eye has a few more tricks to play. Under low light levels the eye can see big objects more clearly than small objects. Increase magnification and you don't improve the contrast but you do enlarge the details to the point that the eye can see them. Keep increasing the magnification and the details will eventually fade as their light is diluted over more and more detector cells. Ultimately, there is optimum balance — sufficient magnification to see details but not so much as to dilute the nebula to invisibility.

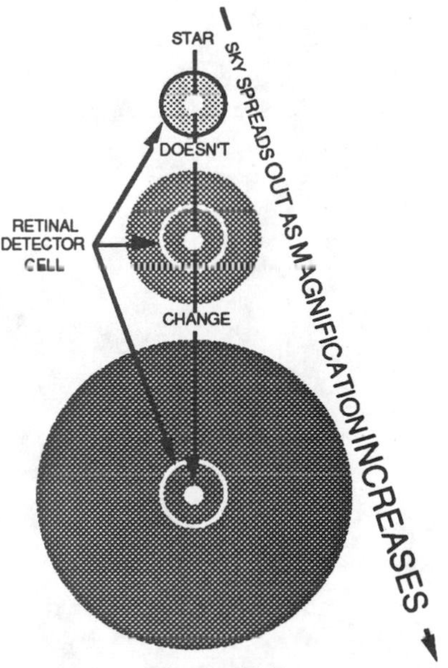

Figure 7-5

The End of the Beginning

Time and practice train your eye even if practice is no more than guess work. Guess the magnitude of each star and then check yourself with The Trained Eye Star Atlas. You'll be surprised how accurate your guesses will become. Guess the magnitude difference and separations among components of multiple star systems. Again you'll surprise yourself. Memorize what you see through the eyepiece, sweep off in some direction, stop and sweep back to your memorized starting point. Guess a bright star's right ascension and declination from looking at the sky before you look it up. Guess the visibility of a star or non-stellar object from its magnitude and dimensions and verify it through the telescope. Look at the colors and magnitudes of multiple stars and visualize their positions on the Hertzsprung-Russell diagram as well as their temperatures, luminosities, sizes, and evolutionary stages. The more you challenge yourself with guessing games, the more you make observing an adventure, the better the observer you become.

Box 7-1: Perception and Deception

Percival Lowell and the Canals of Mars — A Cautionary Tale

"Although skepticism as to the existence of the so-called canals has been now pretty well dispelled ... disbelief still makes a desperate stand "

"It is interesting to recall, in connection with this incredulity about the canals, that precisely the same thing happened in the case of the discovery of Jupiter's satellites and with Huyghens' explanation of Saturn's ring. We are apt to imagine that our age of the world has a monopoly of skepticism. But this is a mistake. The spirit that denies has always been abroad; only in early days he was reputed to be the devil."

Percival Lowell, 1895

In 1877 an Italian astronomer named Schiaparelli first detected on Mars the hint of features that would pique the curiosity of mankind for almost the next 100 years. Schiaparelli glimpsed, through Earth's wavering atmosphere, narrow streaks connecting together the larger dark markings on Mars' surface. Schiaparelli called these streaks *canali* (channels) but thanks to the magic of selective translation, Schiaparelli's "channels" became *canals* in the American press. Channels can be natural, canals never are.

The news of Schiaparelli's canals captured the world's and Percival Lowell's attention. Lowell dedicated his life to astronomy and the idea of Mars as an abode of intelligent life. Yet, despite his decades of work with Mars, marked by excellent science, his canals remained enigmatic.

Enigmatic because of the difficulty of observing such fine markings through the Earth's turbulent atmosphere. The best photographs you can take from the Earth do reveal dark markings on Mars, but none as fine as Lowell recorded. Lowell and his assistants, using their trained eyes during brief moments of perfect seeing, could catch glimpses of detail invisible to the camera. Combining these glimpses, they traced out the gossamer network of canals.

The canals remained an enigma up through the 1960's. Although we have learned a lot more about Mars since Lowell's time (mostly indicating the inhospitablity of Mars to life in general and intelligent life in particular), using Earth based telescopes, Lowell's canals remained enigmatic and irrefutable. Did they indicate real surface features on Mars?

We now know, thanks to extremely detailed photographs of Mars taken from the Viking orbiter, that Lowell's canals are phantoms. So what did Lowell do wrong? Were his canals nothing more than optical illusions? Perhaps Lowell saw in Mars what he most deeply desired — Mars as an abode of intelligent life, revealed by its canals.

Your trained eye lets you create structures as well as discover real ones. The constellations, after all, are mental constructs, not real patterns. You too may create fantasies in the heavens in your pursuit of revealed truth. Give the devil of skepticism his due.

Project 7-1
Star Hopping

Table 1 lists multiple star systems, Table 2 lists non-stellar objects (i.e. clusters, nebulae, galaxies, etc...). The multiple star systems have the usual Bayer-Flamsteed names, (e.g. 94 Aqr). The names beginning with NGC in Table 2 represent New General Catalog non-stellar objects, and names beginning with M are from Messier's catalog.

Your task in this *cloudy*-night exercise is to describe in writing how you are going to find objects selected from Table 1 and 2 using star hopping:

1) Know what you're looking for. Read, understand, and write down the available information so you know what to expect.

2) Pick a bright star as a starting point that is within 30° or so of the object.

3) Pick way stations from the bright star to the object. None fainter than what you're looking for, no more than 10° between each.

4) Use easy slews. Easy slews from one star to another if at all possible. No slew greater than about 10°.

5) Prove you're there. Your predictions of what you should see around the object should match what you see. You need at least two proof objects.

Be sure to include appropriate notes (i.e., a description of what it is you're going to find, the steps you plan to follow to move the telescope to it, where you're going to start, how you will prove you found the right object, etc.). Also record pertinent information on each way station star (i.e. magnitude, right ascension, declination, description of what's around it). Describe the slews you will do to get from one way station to the next (i.e., direction, distance, etc.). To help you, you'll find a star hop route template after the tables.

Tricks of the Trade

1) *If it's easy to find, then find it the easy way. Star hopping is for objects that aren't so easy to find.*

Table 1: Multiple Star Systems

Bayer-Flamsteed Name	RA h	RA m	Dec °	Dec '	Visual Magnitudes				Seperations "		
35 Psc	0	15	8	49	5.9	7.6			11		
η Cas	0	49	57	49	3.5	7.3			12		
ζ Psc	1	14	7	34	5.6	5.8			23.6		
λ Ari	1	58	23	35	4.8	7.1			37.5		
ι Tri	2	12	30	18	5.4	7.0			3.9		
i Cas	2	29	67	24	4.7	7.0	8.3		7.3		
30 Ari	2	37	24	38	6.6	7.3			38.0		
η Per	2	51	55	53	3.9	8.6			8.2		
32 Eri	3	54	-2	58	5.0	6.3			6.9		
λ Ori	5	35	-9	52	3.7	5.6			4.4		
θ Ori	5	35	-5	54	5.0	7.0	7.0	8.0	13	13	17
σ Ori	5	39	-2	36	4.0	7.0	7.0	6.0	13	11	42
ε Mon	6	24	4	34	4.5	6.5			13.2		
δ Gem	7	20	22	00	3.5	8.1			6.1		
ι^1 Cnc	8	47	28	46	4.2.	6.6			30.4		
54 Leo	10	56	24	46	4.5	6.5			6.6		
24 Com	12	35	18	22	5.2	6.7			20.3		
ξ Boo	14	51	19	05	4.8	6.9			7.2		
κ Her	16	08	17	03	5.3	6.5			28		
39 Oph	17	18	-24	17	5.4	6.9			10.8		
95 Her	18	02	21	35	5.2	5.3			6.3		
70 Oph	18	05	2	32	4.2	5.9			2.2		
39 Dra	18	24	58	47	5.1	7.8			3.7		
ε Lyr	18	44	39	40	5.1	6.0	5.1	5.4	207.8	2.7	2.3
15 Aql	19	5	-4	1	5.5	7.2			38.0		
ψ Cyg	19	55	52	25	4.9	7.4			3.1		
γ Del	20	47	16	07	4.3	5.1			9.8		
ξ Cep	22	04	64	37	4.6	6.6			7.6		
94 Aql	23	19	-13	28	5.3	7.5			13.3		
σ Cas	23	59	55	45	5.1	7.2			3.1		

Table 2: Non-Stellar Objects

NGC Number	Messier Number	RA h	RA m	Dec °	Dec '	Visual Magnitude	Type
224	31	0	43	41	16	5.0	G
598	33	1	34	30	39	6.5	G
650-1	76	1	42	51	34	9.0	P
869	h & χ Per	2	19	57	08	4.4	O C
1039	34	2	42	42	46	6.0	O C
	45	3	47	24	07	1.4	O C
7535		4	14	-12	44	9.0	P
1973-1977	42,43	5	36	-4	50	4.0	H II
2168	35	6	09	24	19	5.5	O C
2237-9		6	32	4	37	5.5	H II
2264		6	41	9	53	4.7	O C
2392		7	29	20	55	8.0	P
2422	47	7	37	-14	28	5.0	O C
2548	48	8	14	-5	46	5.5	O C
2632	44	8	40	19	41	4.0	O C
3031	81	9	56	69	03	8.0	G
3034	82	9	56	69	41	9.2	G
4594	104	12	40	-11	37	8.2	G
4736	94	12	51	41	06	8.9	G
4826	64	12	57	21	40	8.6	G
5194	51	13	30	47	11	8.7	G
5272	3	13	42	28	22	6.4	G C
5904	5	15	19	2	05	5.8	G C
6205	13	16	42	36	27	5.9	G C
6341	92	17	17	43	08	6.5	G C
6494	23	17	57	-19	01	7.0	O C
6543		17	59	66	37	8.6	P
6523	8	18	05	-24	19	5.0	H II
24		18	18	-18	25	4.5	O C
6611	16	18	19	-13	46	6.0	H II
6618	17	18	21	-16	10	6.0	H II
6656	22	18	36	-23	55	5.1	G C
	11	18	51	-6	16	6.0	O C
6720	57	18	53	31	01	9.0	P

Table 2 Continued: Non-Stellar Objects

NGC Number	Messier Number	RA h	m	Dec °	'	Visual Magnitude	Type
6826		19	45	50	31	8.8	P
6853	27	19	59	22	53	8.0	P
7009		21	04	-11	22	8.0	P
7078	15	21	30	12	10	6.4	G C
7089	2	21	34	-0	51	6.5	G C
7092	39	21	32	48	26	5.0	O C
7293		22	30	-20	51	6.5	P

Star Hop Template

Object Type ________ Object Name: ________
Right Ascension: ________ Map Number: ________
Declination: ________ Visual Mags: ________ ________
Dimensions: ________ ________ ________
Description:

____
____
____
____
____
____
____

map

Starting Point:
- Names: ________ ________ Visual Magnitude ________
- Right Ascension ________ Declination ________

Way station #1:
- Names: ________ ________ Visual Magnitude ________
- Right Ascension ________ Declination ________
- R.A. Slew to it ________ Dec. Slew to it ________

Way station #2:
- Names: ________ ________ Visual Magnitude ________
- Right Ascension ________ Declination ________
- R.A. Slew to it ________ Dec. Slew to it ________

Way station #3:
- Names: ________ ________ Visual Magnitude ________
- Right Ascension ________ Declination ________
- R.A. Slew to it ________ Dec. Slew to it ________

Destination:
- R.A. Slew to it ________ Dec. Slew to it ________

Proof #1:
- Names: ________ ________ Visual Magnitude ________
- Right Ascension ________ Declination ________
- R.A. Slew to it ________ Dec. Slew to it ________

Proof #2:
- Names: ________ ________ Visual Magnitude ________
- Right Ascension ________ Declination ________
- R.A. Slew to it ________ Dec. Slew to it ________

Star Hop Template

Object Type ________ Object Name: ________

Right Ascension: ________ Map Number: ________

Declination: ________ Visual Mags: ________ ________

Dimensions: ________ ________ ________

Description:

____

____

____

____

____

____

____

map

Starting Point:

Names: ________ ________ Visual Magnitude ________

Right Ascension ________ Declination ________

Way station #1:

Names: ________ ________ Visual Magnitude ________

Right Ascension ________ Declination ________

R.A. Slew to it ________ Dec. Slew to it ________

Way station #2:

Names: ________ ________ Visual Magnitude ________

Right Ascension ________ Declination ________

R.A. Slew to it ________ Dec. Slew to it ________

Way station #3:

Names: ________ ________ Visual Magnitude ________

Right Ascension ________ Declination ________

R.A. Slew to it ________ Dec. Slew to it ________

Destination:

R.A. Slew to it ________ Dec. Slew to it ________

Proof #1:

Names: ________ ________ Visual Magnitude ________

Right Ascension ________ Declination ________

R.A. Slew to it ________ Dec. Slew to it ________

Proof #2:

Names: ________ ________ Visual Magnitude ________

Right Ascension ________ Declination ________

R.A. Slew to it ________ Dec. Slew to it ________

Star Hop Template

Object Type ________ Object Name: ________
Right Ascension: ________ Map Number: ________
Declination: ________ Visual Mags: ________ ________
Dimensions: ________ ________ ________
Description:
____
____
____
____
____
____
____

map

Starting Point:
- Names: ________ ________ Visual Magnitude ________
- Right Ascension ________ Declination ________

Way station #1:
- Names: ________ ________ Visual Magnitude ________
- Right Ascension ________ Declination ________
- R.A. Slew to it ________ Dec. Slew to it ________

Way station #2:
- Names: ________ ________ Visual Magnitude ________
- Right Ascension ________ Declination ________
- R.A. Slew to it ________ Dec. Slew to it ________

Way station #3:
- Names: ________ ________ Visual Magnitude ________
- Right Ascension ________ Declination ________
- R.A. Slew to it ________ Dec. Slew to it ________

Destination:
- R.A. Slew to it ________ Dec. Slew to it ________

Proof #1:
- Names: ________ ________ Visual Magnitude ________
- Right Ascension ________ Declination ________
- R.A. Slew to it ________ Dec. Slew to it ________

Proof #2:
- Names: ________ ________ Visual Magnitude ________
- Right Ascension ________ Declination ________
- R.A. Slew to it ________ Dec. Slew to it ________

Star Hop Template

Object Type ________ Object Name: ________

Right Ascension: ________ Map Number: ________

Declination: ________ Visual Mags: ________ ________

Dimensions: ________ ________ ________

Description:

____

____

____

____

____

____

____

map

Starting Point:

Names: ________ ________ Visual Magnitude ________

Right Ascension ________ Declination ________

Way station #1:

Names: ________ ________ Visual Magnitude ________

Right Ascension ________ Declination ________

R.A. Slew to it ________ Dec. Slew to it ________

Way station #2:

Names: ________ ________ Visual Magnitude ________

Right Ascension ________ Declination ________

R.A. Slew to it ________ Dec. Slew to it ________

Way station #3:

Names: ________ ________ Visual Magnitude ________

Right Ascension ________ Declination ________

R.A. Slew to it ________ Dec. Slew to it ________

Destination:

R.A. Slew to it ________ Dec. Slew to it ________

Proof #1:

Names: ________ ________ Visual Magnitude ________

Right Ascension ________ Declination ________

R.A. Slew to it ________ Dec. Slew to it ________

Proof #2:

Names: ________ ________ Visual Magnitude ________

Right Ascension ________ Declination ________

R.A. Slew to it ________ Dec. Slew to it ________

Star Hop Template

Object Type ________ Object Name: ________

Right Ascension: ________ Map Number: ________

Declination: ________ Visual Mags: ________ ________

Dimensions: ________ ________ ________

Description:

____

____

____

____

____

____

____

map

Starting Point:
- Names: ________ ________ Visual Magnitude ________
- Right Ascension ________ Declination ________

Way station #1:
- Names: ________ ________ Visual Magnitude ________
- Right Ascension ________ Declination ________
- R.A. Slew to it ________ Dec. Slew to it ________

Way station #2:
- Names: ________ ________ Visual Magnitude ________
- Right Ascension ________ Declination ________
- R.A. Slew to it ________ Dec. Slew to it ________

Way station #3:
- Names: ________ ________ Visual Magnitude ________
- Right Ascension ________ Declination ________
- R.A. Slew to it ________ Dec. Slew to it ________

Destination:
- R.A. Slew to it ________ Dec. Slew to it ________

Proof #1:
- Names: ________ ________ Visual Magnitude ________
- Right Ascension ________ Declination ________
- R.A. Slew to it ________ Dec. Slew to it ________

Proof #2:
- Names: ________ ________ Visual Magnitude ________
- Right Ascension ________ Declination ________
- R.A. Slew to it ________ Dec. Slew to it ________

Star Hop Template

Object Type ________ Object Name: ________

Right Ascension: ________ Map Number: ________

Declination: ________ Visual Mags: ________ ________

Dimensions: ________ ________ ________

Description:

____

____

____

____

____

____

____

map

Starting Point:

Names: ________ ________ Visual Magnitude ________

Right Ascension ________ Declination ________

Way station #1:

Names: ________ ________ Visual Magnitude ________

Right Ascension ________ Declination ________

R.A. Slew to it ________ Dec. Slew to it ________

Way station #2:

Names: ________ ________ Visual Magnitude ________

Right Ascension ________ Declination ________

R.A. Slew to it ________ Dec. Slew to it ________

Way station #3:

Names: ________ ________ Visual Magnitude ________

Right Ascension ________ Declination ________

R.A. Slew to it ________ Dec. Slew to it ________

Destination:

R.A. Slew to it ________ Dec. Slew to it ________

Proof #1:

Names: ________ ________ Visual Magnitude ________

Right Ascension ________ Declination ________

R.A. Slew to it ________ Dec. Slew to it ________

Proof #2:

Names: ________ ________ Visual Magnitude ________

Right Ascension ________ Declination ________

R.A. Slew to it ________ Dec. Slew to it ________

Star Hop Template

Object Type __________ Object Name: __________
Right Ascension: __________ Map Number: __________
Declination: __________ Visual Mags: __________ __________
Dimensions: __________ __________ __________
Description:

map

Starting Point:
Names: __________ __________ Visual Magnitude __________
Right Ascension __________ Declination __________
Way station #1:
Names: __________ __________ Visual Magnitude __________
Right Ascension __________ Declination __________
R.A. Slew to it __________ Dec. Slew to it __________
Way station #2:
Names: __________ __________ Visual Magnitude __________
Right Ascension __________ Declination __________
R.A. Slew to it __________ Dec. Slew to it __________
Way station #3:
Names: __________ __________ Visual Magnitude __________
Right Ascension __________ Declination __________
R.A. Slew to it __________ Dec. Slew to it __________
Destination:
R.A. Slew to it __________ Dec. Slew to it __________
Proof #1:
Names: __________ __________ Visual Magnitude __________
Right Ascension __________ Declination __________
R.A. Slew to it __________ Dec. Slew to it __________
Proof #2:
Names: __________ __________ Visual Magnitude __________
Right Ascension __________ Declination __________
R.A. Slew to it __________ Dec. Slew to it __________

Star Hop Template

Object Type ______ Object Name: ______
Right Ascension: ______ Map Number: ______
Declination: ______ Visual Mags: ______ ______
Dimensions: ______ ______ ______
Description:

______
______
______
______
______
______
______

map

Starting Point:
Names: ______ ______ Visual Magnitude ______
Right Ascension ______ Declination ______
Way station #1:
Names: ______ ______ Visual Magnitude ______
Right Ascension ______ Declination ______
R.A. Slew to it ______ Dec. Slew to it ______
Way station #2:
Names: ______ ______ Visual Magnitude ______
Right Ascension ______ Declination ______
R.A. Slew to it ______ Dec. Slew to it ______
Way station #3:
Names: ______ ______ Visual Magnitude ______
Right Ascension ______ Declination ______
R.A. Slew to it ______ Dec. Slew to it ______
Destination:
R.A. Slew to it ______ Dec. Slew to it ______
Proof #1:
Names: ______ ______ Visual Magnitude ______
Right Ascension ______ Declination ______
R.A. Slew to it ______ Dec. Slew to it ______
Proof #2:
Names: ______ ______ Visual Magnitude ______
Right Ascension ______ Declination ______
R.A. Slew to it ______ Dec. Slew to it ______

Project 7-2
Celestial Sights

In Project 7-1, you practiced planning an observing session and laying out a star hop route to find faint objects. Now you get to put that knowledge to use. In addition, you should continue to practice the techniques you learned in Chapter 4 to reinforce your control over the telescope:

1) Step back and look where the telescope is pointed. If you're supposed to be in Orion, your telescope should be pointed towards Orion.

2) Go back to your starting point and check your finder's alignment.

3) Go back to your starting point and check it's coordinates to what the right ascension and declination dials display.

   A) If the right ascension is wrong, correct it by dialing up the correct right ascension.

   B) If the declination dial is way off (more than 3°), then go back to square zero and redo the polar axis alignment (Chapter 4).

   C) If the declination dial is slightly misaligned (less than 3°), then just compute and apply the *declination error.*

If you can't find any of the objects, you should explain why not. If you think it's below the horizon, remember how to prove it (like you did for Project 4-2). If you can't resolve a multiple star, try and try again. An observation template is included at the end of this project.

Tricks of the Trade (repeated from Project 5-2 for your convenience):

1) The trick to determining NESW as seen in the eyepiece is to let the star's motion tell you which way is west.

   A) Center the star in the eyepiece.

   B) Turn off the clock drive.

   C) The stars will drift to the west side of the eyepiece.

D) Put your finger on the side of the eyepiece the stars moved to. That is the west side of the eyepiece.

E) Label that side of your sketch using W for the west.

F) Restart your clock drive so that you don't get too far behind and check the right ascension.

G) On your sketch, east is directly opposite west. Determining north or south depends on whether or not you are using a diagonal eyepiece holder.

If you are not using a diagonal, then south is 90 degrees counter-clockwise from east, and north is 90° counter-clockwise from west.

If you are using a diagonal, then south is 90° clockwise from east, and north is 90° clockwise from west.

2) *Always use a low-power eyepiece to acquire and center the star in the telescope. In your writeup, explain the advantage of using low power for acquisition.*

3) *Develop a sense for how bright stars of different magnitude look through different eyepieces and the finder.*

4) *Develop a sense of how different separations for multiple stars look through different eyepieces and the finder.*

5) *To see color in bright stars, defocus the images slightly.*

6) *Check the Multiple Star System pages of The Trained Eye Star Atlas for additional information on each multiple star.*

7) *For multiple stars, the rule of thumb for selecting an eyepiece is to use an eyepiece with a focal length just equal to or less than the separation of the components. That is, if the separation is 20", then a 20 mm eyepiece will just let you see both, a 10 mm eyepiece will let you see them easily.This eyepiece selection rule assumes novice eyes, an f-ratio of 6 to 8, and objective diameter of 6 to 10 inches. It also assumes that both components are of equal magnitude. If they are 5 magnitudes or more different, you'll need even more magnification.*

8) *Always check your dials before leaving a star. If the right ascension dial is wrong, correct it. If the declination dial is wrong, calculate the Dec Error.*

9) *Start with the multiple star systems until you get the hang of what you're doing.*

10) *Start with things lowest in the west since they will be gone in a few weeks.*

11) *Help yourself by writing down problems in your notes as you encounter them.*

Multiple Star Questions

Include your observations from Project 4-3 in answering these questions.

1) What environmental factors do you think might affect visibility (e.g., altitude, seeing, transparency, etc.) ? Explain the effect of each.

   Hint:: Give examples from your observations.

2) Was there a systematic difference between the temperatures of the brightest component of a multiple star and the temperatures of the faintest component stars? Describe.

   Hint: The colors of stars range from blue hot to red cool:

   blue blue-white white yellowish-white light -yellow deep-yellow orange red

   Hint: List the number of stars of each color for the brightest component and for the faintest components of a multiple star system. Color categories are: blue, blue-white, yellowish-white, light yellow, yellow, orange, red, no color. For faint stars white means no color.

3) How many stars showed color?

   Hint: Remember that for faint stars white means NO color so don't count faint white stars.

4) What was the most common star color?

   Hint: See Hint for question 2.

5) What was the magnitude of the faintest component star of a multiple star in which you saw color? What was that color?

6) Which multiple star had the component star with the coolest temperature? Was it the brighter or fainter component star?

7) Use the colors and magnitude differences of the multiple stars to place their components on the H-R diagram. For those where only the brighter component star shows color, use the H-R diagram to deduce the color of the fainter component star.

   Hint: When both stars show color:

   1) Draw a vertical line on the H-R diagram at the color of each (each star's color line).
   2) PLace the brighter star on its color line at the top the H-R diagram .
   3) Place the fainter on its color line below the brighter star by the amount of the magnitude difference between the two stars.

4) *Slide the two stars down their color lines, maintaining their vertical seperation, until they align with features of the H-R diagram (i.e. cool giant, main sequence, white dwarf).*

When only the brighter star shows color:

1) *Draw a vertical line on the H-R diagram at the color of the brighter component star (its color line).*
2) *Place the brighter star on its color line at the top of the H-R diagram (the bead).*
3) *Place a horizontal line straight across the H-R diagram below the brighter star by the amount of the magnitude difference between the two component stars.*
3) *Slide the two stars (bead and horizontal line) down the H-R diagram until the brighter star aligns with a feature of the H-R diagram (i.e. cool giant or main sequence) .*
5) *Where the horizontal line (representing the fainter component star) crosses a feature of the H-R diagram (I.e. Main sequence, white dwarf) identifies its color.*

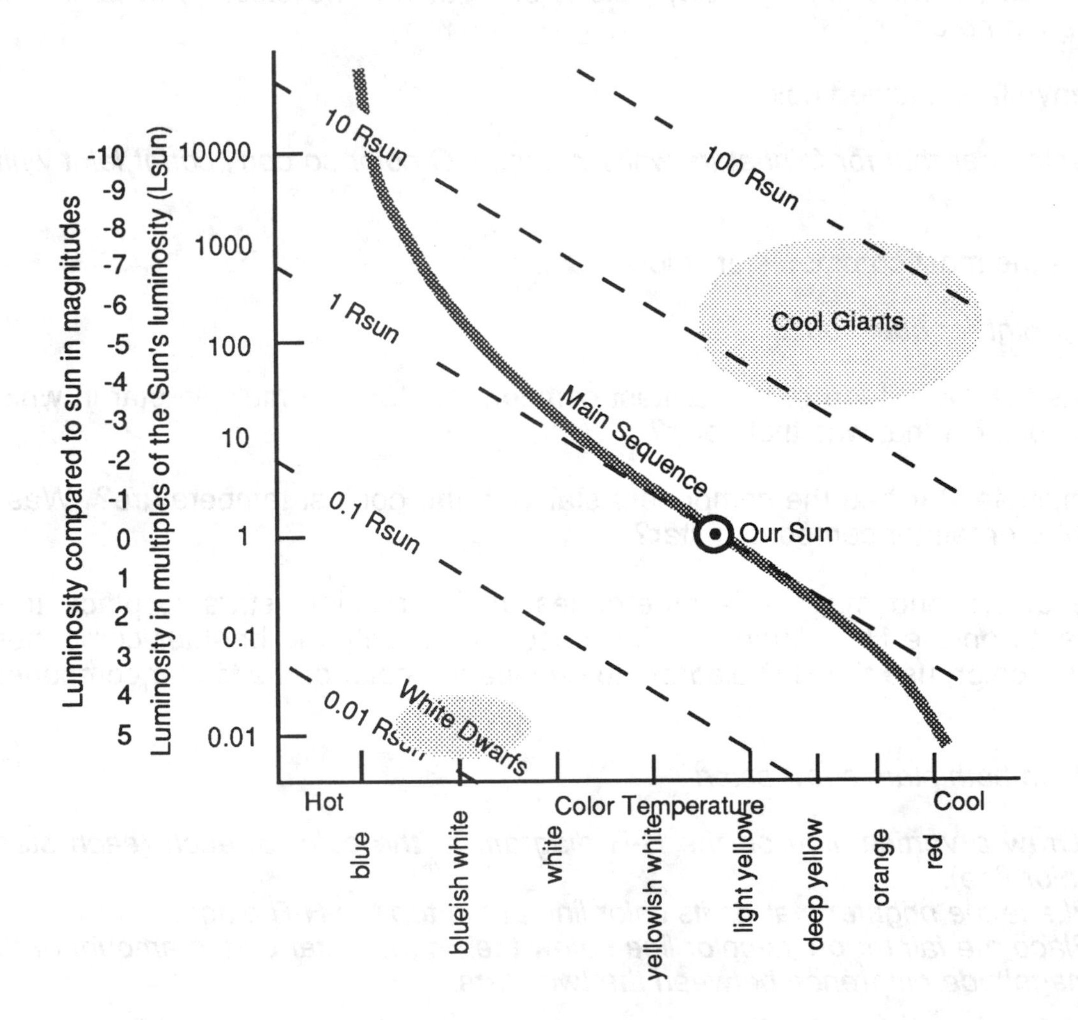

Label the stars on the sketch with the star name followed by the letter A, B, C or D from brightest component to faintest. For example, the two components of γ Del would be labeled γ Del A and γ Del B.

8) List the number of white dwarfs, main sequence, and giant stars.

9) Which multiple star(s) had the biggest star?

10) Which multiple star(s) had the youngest star?

11) Which multiple star(s) had the most evolved star?

Non-Stellar Object Questions

12) Which non-stellar objects (open star clusters, globular clusters, HII regions, planetary nebulae, galaxies) showed color?

13) Did any of the non-stellar objects (star clusters, globular clusters, HII regions, planetary nebulae, galaxies) resolve into stars? What were they and how much magnification did they require?

14) How did the visibility of each non-stellar object (open star clusters, globular clusters, HII regions, planetary nebulae, galaxies) change with magnification? Which eyepiece gave the best visibility for each type?

15) Which type of non-stellar objects, if any, did you find in close proximity to HII regions? What were the colors of stars associated with HII regions?

16) What are the relative distances of the star clusters you observed?

Hint: Only compare open clusters to open clusters and globular clusters to globular clusters. Judging the angular size of a cluster will be tricky since they have no sharp edges, they just fade into the background stars. Still, using the law of perspective, the clusters with the largest angular size is closest, a cluster with half that angular size is twice as distant, etc.... — assuming of course that all clusters have the same real size.

OBJECT

ID	RA
Type	Dec
Vis Mag	Epoch
Dimension	

TELESCOPE

Type	Eyepiece
Aperture	Power
F-Ratio	Filter
Focal Len	

ENVIRONMENT

Where	Latitude
Date	Longitude
Time	Altitude
Seeing	Azimuth
Transparency	Conditions
Visibility	

NOTES

OBJECT

ID	RA
Type	Dec
Vis Mag	Epoch
Dimension	

TELESCOPE

Type	Eyepiece
Aperture	Power
F-Ratio	Filter
Focal Len	

ENVIRONMENT

Where	Latitude
Date	Longitude
Time	Altitude
Seeing	Azimuth
Transparency	Conditions
Visibility	

NOTES

OBJECT

ID
Type
Vis Mag
Dimension

RA
Dec
Epoch

TELESCOPE

Type
Aperture
F-Ratio
Focal Len

Eyepiece
Power
Filter

ENVIRONMENT

Where
Date
Time
Seeing
Transparency
Visibility

Latitude
Longitude
Altitude
Azimuth
Conditions

NOTES

OBJECT

ID
Type
Vis Mag
Dimension

RA
Dec
Epoch

TELESCOPE

Type
Aperture
F-Ratio
Focal Len

Eyepiece
Power
Filter

ENVIRONMENT

Where
Date
Time
Seeing
Transparency
Visibility

Latitude
Longitude
Altitude
Azimuth
Conditions

NOTES

OBJECT

ID
Type
Vis Mag
Dimension

RA
Dec
Epoch

TELESCOPE

Type
Aperture
F-Ratio
Focal Len

Eyepiece
Power
Filter

ENVIRONMENT

Where
Date
Time
Seeing
Transparency
Visibility

Latitude
Longitude
Altitude
Azimuth
Conditions

NOTES

OBJECT

ID
Type
Vis Mag
Dimension

RA
Dec
Epoch

TELESCOPE

Type
Aperture
F-Ratio
Focal Len

Eyepiece
Power
Filter

ENVIRONMENT

Where
Date
Time
Seeing
Transparency
Visibility

Latitude
Longitude
Altitude
Azimuth
Conditions

NOTES

OBJECT

ID	RA
Type	Dec
Vis Mag	Epoch
Dimension	

TELESCOPE

Type	Eyepiece
Aperture	Power
F-Ratio	Filter
Focal Len	

ENVIRONMENT

Where	Latitude
Date	Longitude
Time	Altitude
Seeing	Azimuth
Transparency	Conditions
Visibility	

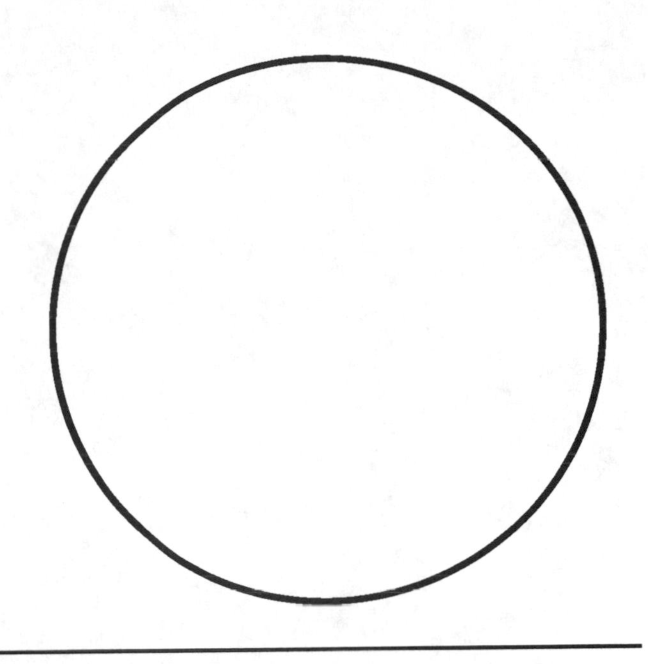

NOTES

OBJECT

ID	RA
Type	Dec
Vis Mag	Epoch
Dimension	

TELESCOPE

Type	Eyepiece
Aperture	Power
F-Ratio	Filter
Focal Len	

ENVIRONMENT

Where	Latitude
Date	Longitude
Time	Altitude
Seeing	Azimuth
Transparency	Conditions
Visibility	

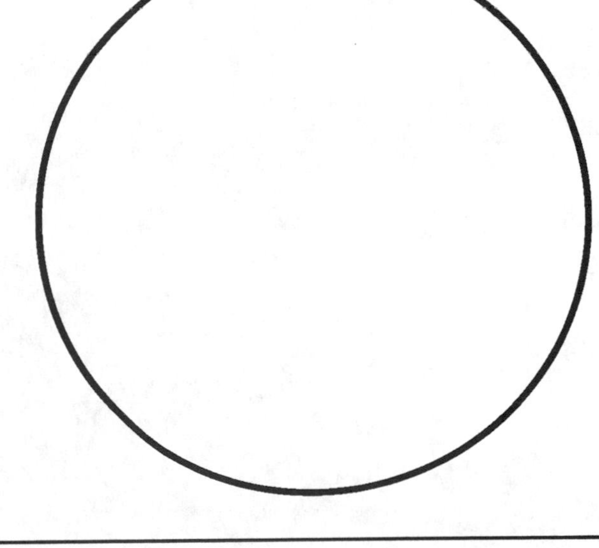

NOTES

Appendix A
Additional Projects

PROJECT A-1
Waiting for the Sun

The origin of astronomy is not "shrouded in the mists of time." Although astronomy has originated many times in many places, it has always originated for a clear reason; the discovery of the harmony between the celestial cycles of the Sun (and the Moon) and the rhythms of life on Earth. Observe the Sun's cycle and discover the year of the stars — the year of the seasons. The observations required are simple and enjoyable. All you have do is watch the sunset.

First remember what happens to the time of sunset over the next few months. Then pick a place where you can be at sunset and see the horizon from the southwest to the northwest (from azimuth 225° to azimuth 315°). If you're an early bird, you may want to watch the sunrises instead, so adjust your location and horizon accordingly (azimuth 45° to 135°). If so pick a place where you can see the horizon and make a sketch on a sheet of paper of your horizon similar to the following:

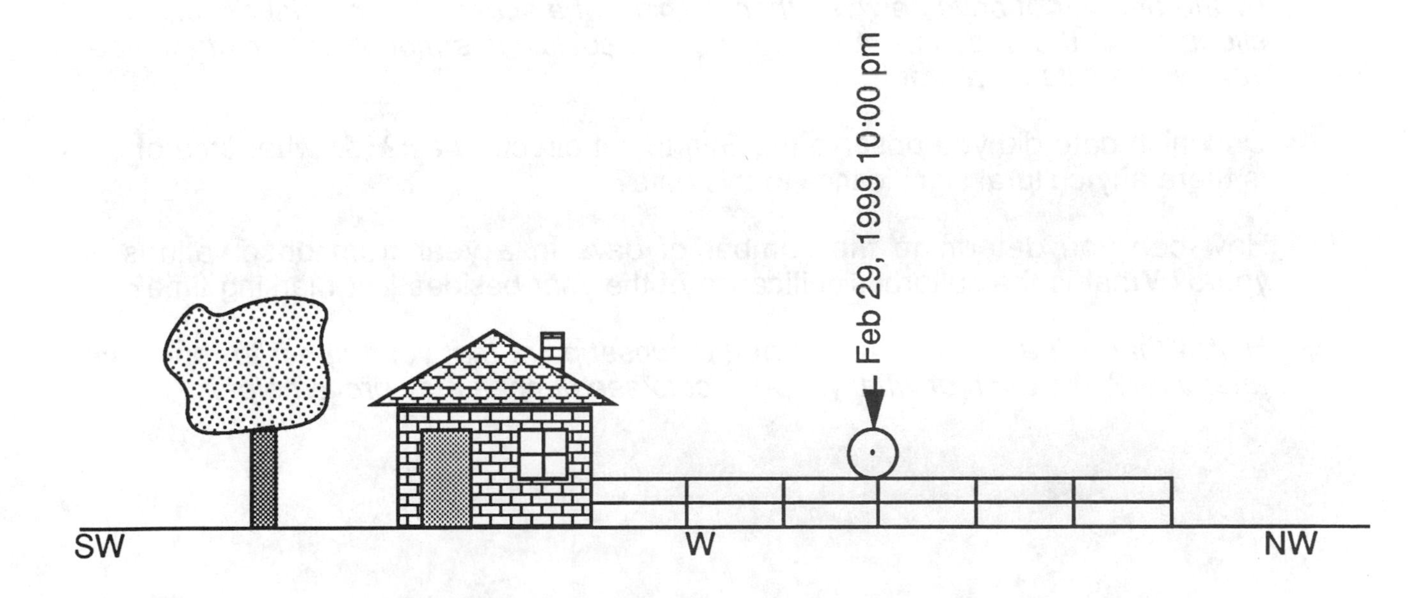

Include prominent stationary features such as buildings, trees, etc. Do not include moveable objects such as cars, animals or small children. Remember you must stand in the same place each time, so mark your spot carefully.

Watch the sunset once a week. Mark on your sketch where the Sun touches the horizon and note the date and time of each observation as shown above. If you can't see exactly where it sets because of buildings, fog, clouds, etc., sketch where it disappears and make your best guess of where it will set on the horizon.

For one of your sunset observations, mark the position of the Sun (as best you can) every 15 minutes for the last hour before sunset.

Questions

Answer these questions after you have completed your series of observations. Ignore the week to week variations since they are dominated by how well you observe not by the what the Sun is doing. Look instead for the month to month trends.

Explain how your observations support your answers to each question. Once you've written your explanation, read it while looking at your observations. If the observations don't match your explanation, then try again.

1) Does the Sun set straight down?

2) Which way does the point of sunset slide along the horizon and does it always slide in the same direction?

3) Does the point of sunset slide at a constant speed along the horizon or does it speed up/slow down?

   Hint: Measure the distance between successive sunsets in millimeters and divide by the number of days between them. This is the speed of the point of sunset as it slides along the horizon. A handy way to visualize this motion is to plot the speed versus the date on a graph.

4) On which date did you observe the Sun to set directly west? At what time of day? Is there any cultural significance to this date?

5) How can you determine the number of days in a year from observations like yours? What is the cultural significance of the year besides just marking time?

   Hint: What is it about the motion of the sunset point that you expect to repeat year after year? A sketch of what you expect to see happen may prove helpful.

PROJECT A-2
A Month of Moon

The Moon establishes another major cycle in the heavens and the calendar; the month. Its ever changing appearance has intrigued astronomers, poets and lovers for millenia. Sometimes it is a brash shining disk casting shadows in the night; at other times, just a thin sliver of a smile. Always it is in a different part of the sky. The rhyme and reason to this mystery are easily solved by watching.

Observe the Moon once a day for 30 days. Start with the first quarter Moon in the evening sky. Put the following information in your notes for each observation:

1) A sketch of the Moon, showing its phase, correctly oriented with the horizon like the following:

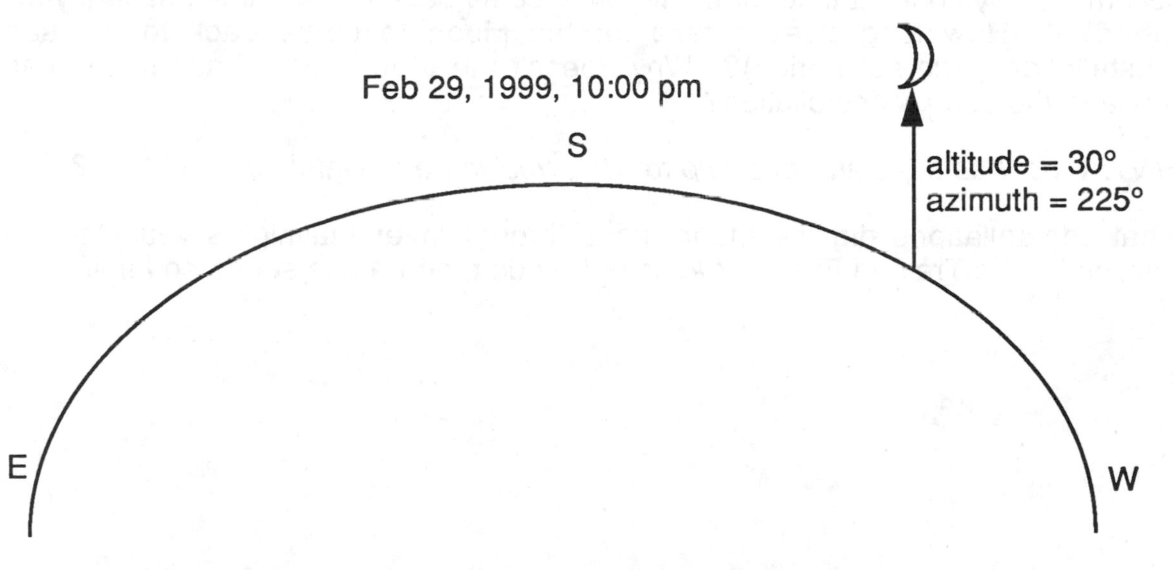

2) Date and time of day of your observation, altitude and azimuth of the Moon.

3) If possible, the distance from the Sun to the Moon in hands (since you know each hand is 10°, you can convert the number of hands to the number of degrees).

4) For the first four nights, note which constellation the Moon is in and mark its position on the appropriate star map in The Trained Eye Star Atlas. Repeat this for the last four nights.

Hint: Measure the distance from the Moon to three bright stars surrounding the Moon in hands. Multiply the number of hands by ten and you have the number of degrees from each star to the Moon. Draw a circle in The Trained Eye Star Atlas that number of degrees in radius around each star. You can use the declination scale on the side of the map to set your compass. Where all three circles merge (or come closest together) is the location of the Moon.

Questions

Answer the following questions after you have made your month's worth of observations. Explain how your observations support your answers to each question.

1) Was the Moon ever visible during the daylight hours? Describe when, where, appearance, distance from the Sun.

2) Were you ever able to see the dark side of the Moon? Describe when, where, appearance, distance from the Sun.

3) How does the phase depend upon the distance (in hands) of the Moon from the Sun? That is, how does the amount of illuminated Moon compare to the number of degrees it is away from the Sun?

4) Which way does the Moon move relative to the stars (relative to the constellations)?

5) How many days does it take for the Moon to come back to the same phase (synodic period) ? How long does it take for the Moon to come back to the same constellation (sidereal period)? Why doesn't the Moon come back to the same phase in the same constellation?

   HINT: What has the Earth been up to while you've been watching the Moon?

6) What constellations did the Moon move through over the nights you plotted its position in The Trained Eye Star Atlas? Why do their names sound so familiar?

Project A-3
Wheels Within Wheels

Mars and Venus provide the most spectacular examples of the motion of the planets relative to the stars. Mars moves, on average, about 1/2° per night (the width of the full Moon), while Venus can sprint along at 2° per night. Because they move so fast, you can actually compare what you see on any given night to your memory of its position the night before — you can see their motions. They generally drift *prograde* (eastward) but occasionally will slow to a *standstill* , reverse direction to move *retrograde* (westward) for a short while, once again slow to a *standstill*, turn and resume their *prograde* progress. The complexities of their apparent motions confounded Ptolemey and inspired Copernicus; Ptolemy's model of the universe was as complex as the motions, Copernicus' model modified with Kepler's ellipses revealed the underlying simplicity of the solar system. There are other aspects of the planets you can observe as they perform their dance among the stars; i.e. visual magnitude and appearance.

Part A : Visual Observations

Observe Mars and/or Venus depending upon which is available at the time of night that you're awake. Once a night, at the same time each night, mark the position of Mars and/or Venus in pencil on the appropriate star map in The Trained Eye Star Atlas. Use the following symbols for the planets on your star map:

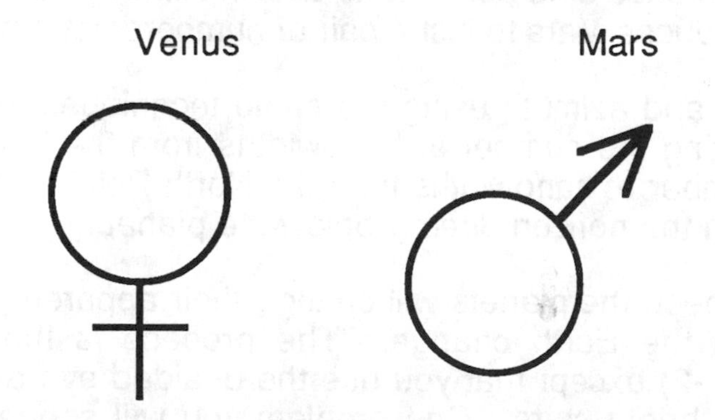

Keep track of each symbol by writing the observation number inside the circle. This number connects the symbol to Table 1 (at the end of the project) where you record date and time of your observation, the right ascension and declination that you interpolate off of The Trained Eye Star Atlas or equatorial mounting, and the altitude and azimuth you measure with handwidths. Below is an example of what your star map should look like:

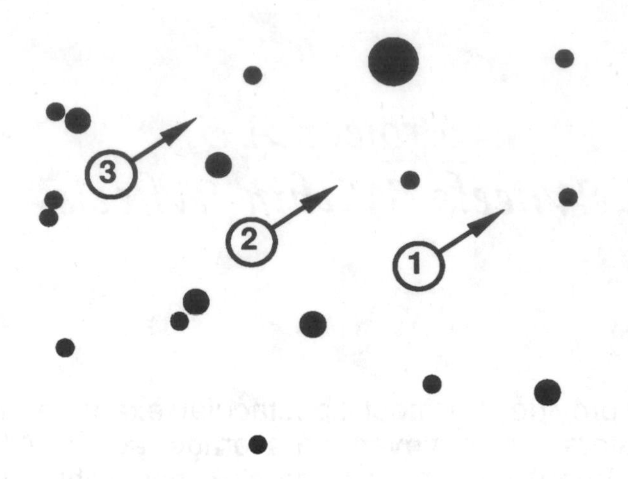

There are three techniques you should use to measure the right ascension and declination of the planets among the constellations:

1) You can get a quick estimate of its position by measuring the planet's distance from nearby bright stars in the sky in handwidths (each hand is 10°). You can convert handwidths to degrees to measure the planet's position from the same bright stars on your star map and mark its position. This least accurate technique is good to about ±1/2 hand (5°) but fixes the observation in your memory.

2) If you can recognize quite a few nearby faint stars, you may be able just to look at its position among the faint stars and mark its position in The Trained Eye Star Atlas to match. With this technique, you should be able to get close to ±1° accuracy in your table and fix the observation in your memory just as well.

3) If you have a telescope with an aligned equatorial mounting, you can just point the telescope at the planet and read off the declination and right ascension directly and you should be accurate to ±0.5°. Although this is the most accurate technique, it reduces Mars to just a pair of numbers not a mental image.

Measure altitude and azimuth using the same techniques as you used in Project 2-1 and 2-2 by counting the number of handwidths from the horizon up to the planet (altitude) and the number of handwidths from the North Point on the horizon clockwise around to the point on the horizon directly below the planet.

As you might expect, the planets will change their apparent visual magnitudes as their distances from the Earth change. The process is the same as for visual photometry (Project 4-2) except that you use the unaided eye and compare its visual magnitude to nearby bright stars. One problem you will sooner or later hit is when Mars' grows brighter than all the stars you have to compare to it. However, you can cut Mars' glow back down to size with a pair of sunglasses.

Your sunglasses will diminish Mars' glare sufficiently to compare to nearby stars observed without sunglasses. Obviously Mars' apparent visual magnitude will be fainter depending upon how many magnitudes of light your sunglasses absorb.

Fortunately, you can easily measure how many magnitudes your sunglasses absorb and apply that amount to get the correct visual magnitude for Mars:

Apparent Magnitude = Apparent Magnitude With Sunglasses - Sunglass Correction

There are a couple of ways to measure how many magnitudes your sunglasses absorb:

1) Observe the brightest star of known magnitude through sunglasses and compare it to other not quite as bright stars of known magnitude observed without sunglasses until you get a match. Then:

   Sunglasses Correction = Magnitude of Faint Star - Magnitude of Bright Star

2) Obtain two small battery powered lights that look like same brightness at the same distance. Place one about 10 feet away. While observing it with your sunglasses, have a partner move the other light further and further away until, observing it without sunglasses, it looks the same brightness as the nearby one observed with sunglasses. Measure the distance to both lights and calculate the sunglass correction:

$$\text{Sunglass Correction} = 5 * \log_{10}\left(\frac{\text{distance to light viewed without sunglasses}}{\text{distance to light viewed with sunglasses}}\right)$$

After several months (possibly even after several weeks) you can start to answer the questions, but be careful: your answers may change the longer you watch.

Part A Questions

1) Did Venus and/or Mars move relative to the stars?

2) Which constellations did Venus and/or Mars move through and in which directions? What's special about these constellations?

   Hint: Which constellations does the Sun appear to move through?

3) Did Venus and/or Mars ever change direction and speed? If they did, then when and in which constellations?

4) How have the distance and direction from Venus/Mars to the Sun in the sky varied over the time you've been observing and how has that related to the apparent motion of Venus and/or Mars among the constellations?

   Hint: You've been observing the planet at the same solar time each night and so you know which horizon the Sun is below (eastern or western) — you roughly know its altitude (negative) and azimuth (around either 90° or 270°). Plot all your measurements of altitude and azimuth on a graph, and label them in with the direction the planet was moving among the constellations (prograde, retrograde, standstill). Also plot the Sun's position below whatever horizon it is nearest (western horizon for early evening, eastern horizon for early morning). From this graph you can roughly measure the distance of Venus and/or Mars to the Sun with a ruler and relate the motion of the planet (prograde, retrograde, standstill) to its distance (in degrees) and direction from the Sun (if the Sun is below the western horizon it is west of the planet, if it is just below the eastern horizon, it is east of the planet).

5) **When in its motion was Mars the brightest? Which way was Mars moving (prograde, standstill, retrograde) when it reached its brightest visual magnitude?**

Table 1: Planet Observations

Obs. Num.	Date	Time	RA		Dec		Alt	Azi	Visual Mag.	Diam.
			h	m	°	'	°	°		"

Part B: Telescopic Observations

Sketch the telescopic appearance of Venus and/or Mars on a copy of the Observation Sheet (at the end of this project) each time you measure its position. Pay special attention to getting N, E, S and W correct. Measure the size of each planet as follows:

1) Find a guiding eyepiece — a high power eyepiece with a crosshair generally used to guide the telescope during long time exposure astrophotography.

2) Install the guiding eyepiece with one of the crosshairs aligned N-S and the other E-W.

3) Turn off the clock drive and time how long the planet takes to drift past the N-S crosshair in seconds.

4) Convert the drift time (seconds) to arc-seconds (") using the following relationship:

1 second of time = 15" (arc-seconds)

After several months (possibly even after several weeks), you can start to answer the questions but be careful: your answers may change the longer you watch.

Part B Questions

6) Did Venus change appearance dramatically? Relate its appearance to its position in its orbit relative to the Earth and Sun.

Hint:

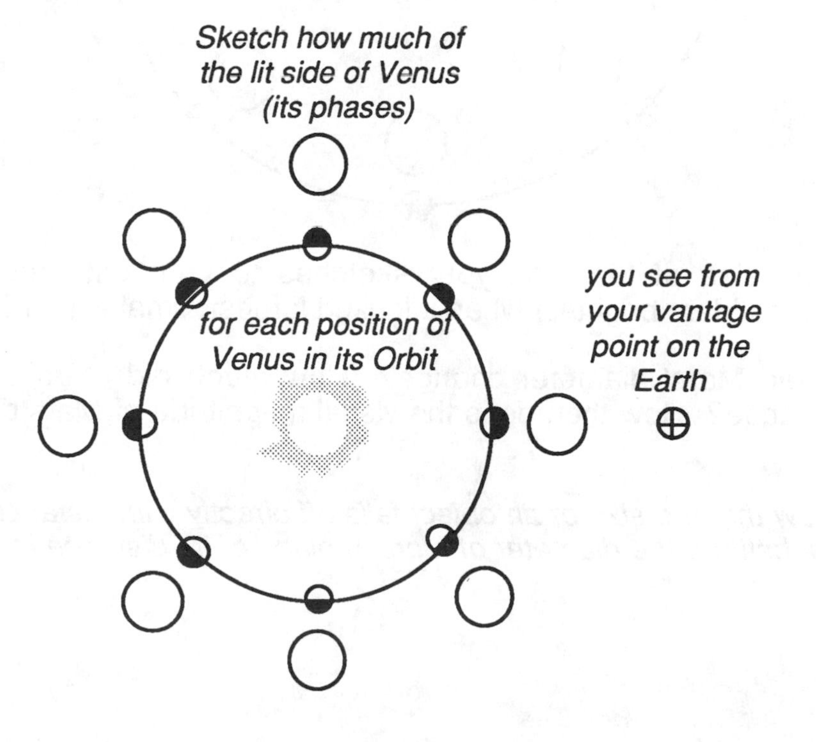

7) At which phase does Venus look the biggest (i.e., new, crescent, quarter, full)? At which phase does Venus look the smallest? Therefore, at which phase is Venus closest? Why does this make sense?

Hint: Relate to your drawing in question 6.

8) What was the phase of Venus when it was at its brightest and faintest? What was the size of Venus at its brightest and faintest? Explain any discrepancy.

Hint: Consider not only the inverse square law, but also the phase of Venus.

9) Does Mars show phases? Explain.

Hint:

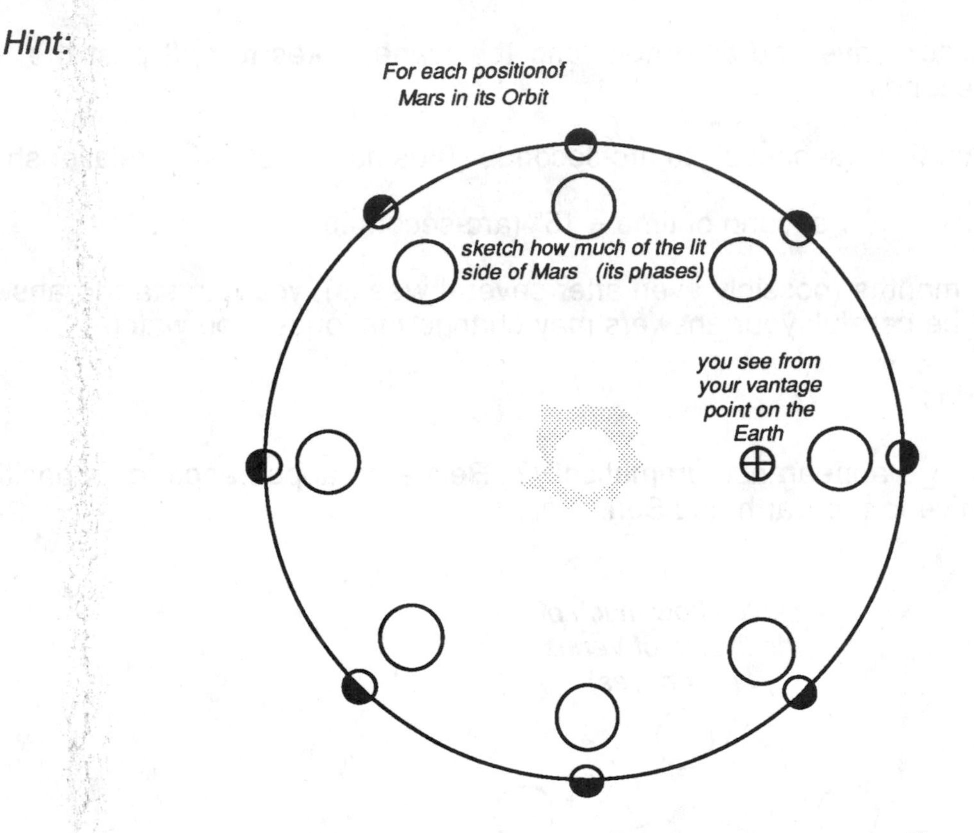

10) Compare the size of Mars on your sketches to your estimates of its visual magnitude. Was Mars brightest when it looked largest, smallest, or in between?

11) How much did Mars' diameter change? How much did Mars' apparent visual magnitude change? How then does the visual magnitude of Mars depend upon its distance?

Hint: You know that the size of an object falls off directly with distance (Chapter 5) so you can substitute the diameter of Mars in place of its distance in the inverse square law;

$$\frac{\text{brighter apparent brightness}}{\text{fainter apparent brightness}} = \left(\frac{\text{larger diameter}^2}{\text{smaller diameter}^2}\right)$$

The left side of this equation is just a brightness ratio which you can compute from the magnitude difference using the following table:

Magnitude Difference	Brightness Ratio
-2.5	10
-2.4	9.1
-2.3	8.3
-2.2	7.6
-2.1	6.9
-2.0	6.3
-1.9	5.7
-1.8	5.2
-1.7	4.8
-1.6	4.4
-1.5	4
-1.4	3.6
-1.3	3.3
-1.2	3
-1.1	2.7
-1.0	2.5
-0.9	2.3
-0.8	2.1
-0.7	1.9
-0.6	1.7
-0.5	1.6
-0.4	1.4
-0.3	1.3
-0.2	1.2
-0.1	1.1
0.0	1.0

To prove the inverse square law just plug your largest and smallest measurements of the size of Mars and the associated visual magnitudes (converted to a brightness ratio the above table); you may want to average several observations together to improve accuracy). If both sides of the equation come out about equal, then you've validated the inverse square law.

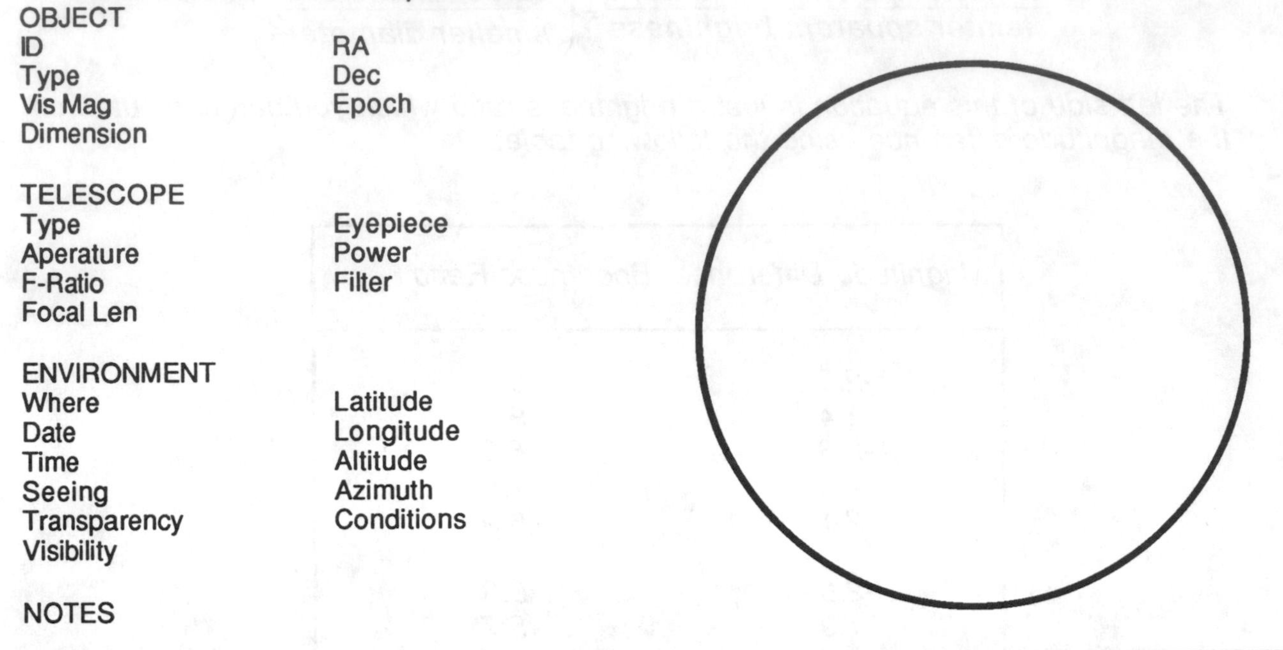

OBJECT
ID
Type
Vis Mag
Dimension

RA
Dec
Epoch

TELESCOPE
Type
Aperature
F-Ratio
Focal Len

Eyepiece
Power
Filter

ENVIRONMENT
Where
Date
Time
Seeing
Transparency
Visibility

Latitude
Longitude
Altitude
Azimuth
Conditions

NOTES

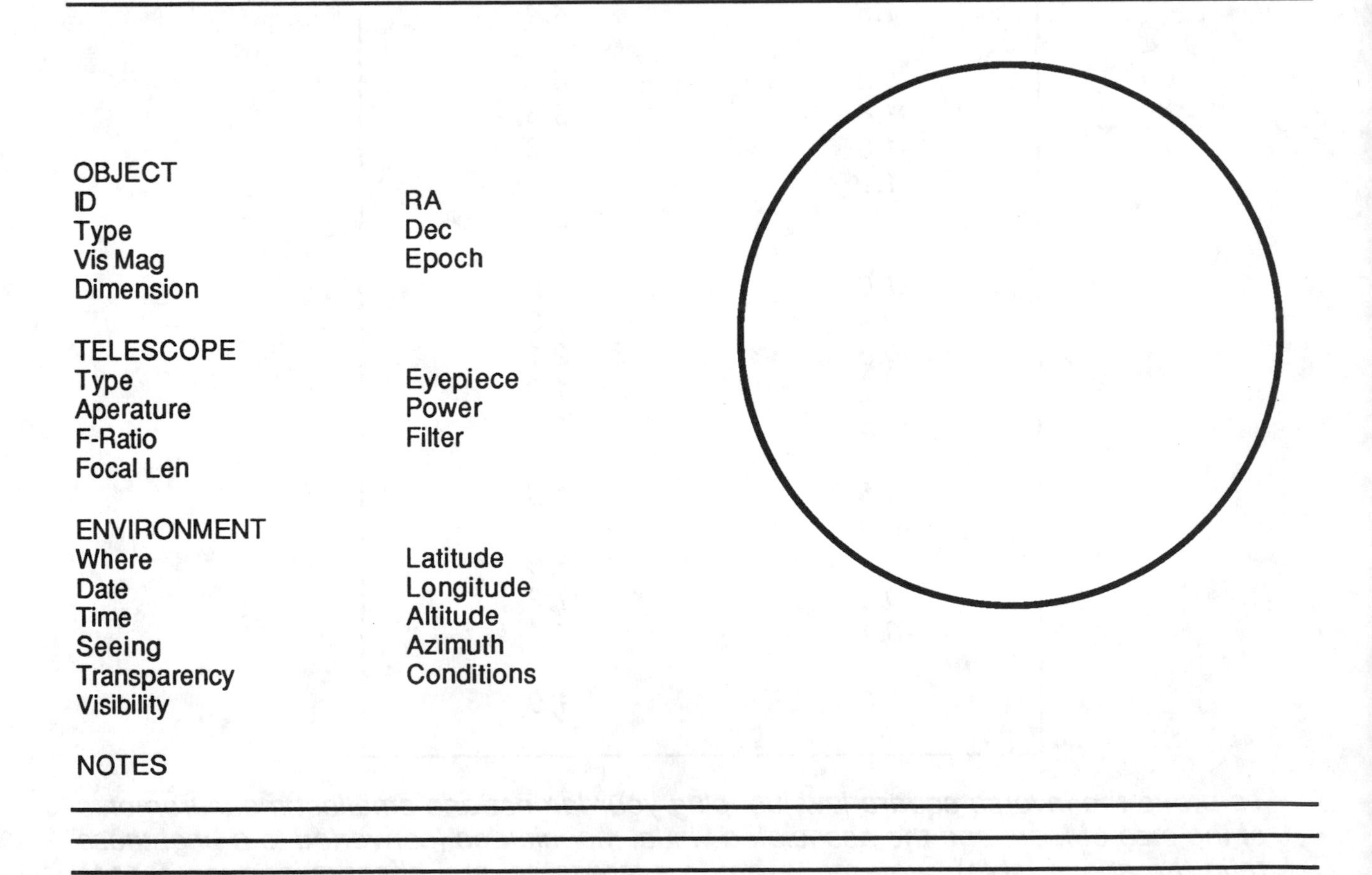

OBJECT
ID
Type
Vis Mag
Dimension

RA
Dec
Epoch

TELESCOPE
Type
Aperature
F-Ratio
Focal Len

Eyepiece
Power
Filter

ENVIRONMENT
Where
Date
Time
Seeing
Transparency
Visibility

Latitude
Longitude
Altitude
Azimuth
Conditions

NOTES

Project A-4
The Shape of the Galaxy

Copernicus spun the Earth away from the center of the universe into circles about the Sun. Controversial as this was, Copernicus' *heliocentric* model of the universe kept a fair dose of egocentricism; the universe revolved about our Sun. Giordano Bruno displaced ego even further by proposing that a universe full of Suns, each with a family of planets inhabited by thinking beings, each and every one of which believed that the universe revolved about he/she/it.

What Bruno called the universe we now know call our galaxy — the two words diverged in meaning less than 100 years ago. You cannot answer the bigger question about the structure of the universe and our place within it because your eyes are limited to only the closest and brightest components of the universe — those stars lying within our own galaxy. You can learn something about the structure of our galaxy by looking at the distribution of stars around you. The techniques you learn by studying the galaxy with your unaided eyes are the same techniques astronomers, with their telescopically aided eyes, use to study the structure of the larger universe and our place within it.

Your Turn

Start by thinking about what structure the galaxy *might* have, how *might* stars be distributed in space. Two possible models that come quickest to mind are a spherical distribution of stars and a flattened one. With model in mind, you can ask yourself, "How would this galaxy look from the inside?" Take the spherical galaxy first. If you were at the center you would see stars evenly distributed around you in all directions — if you were near the edge, you'd see more stars in the hemisphere of sky towards the center than looking away from the center. From the center of the flattened galaxy, you'd see stars in a even band around your head, while if you were near the edge you'd see stars in an uneven band around your head, concentrated back towards the center.

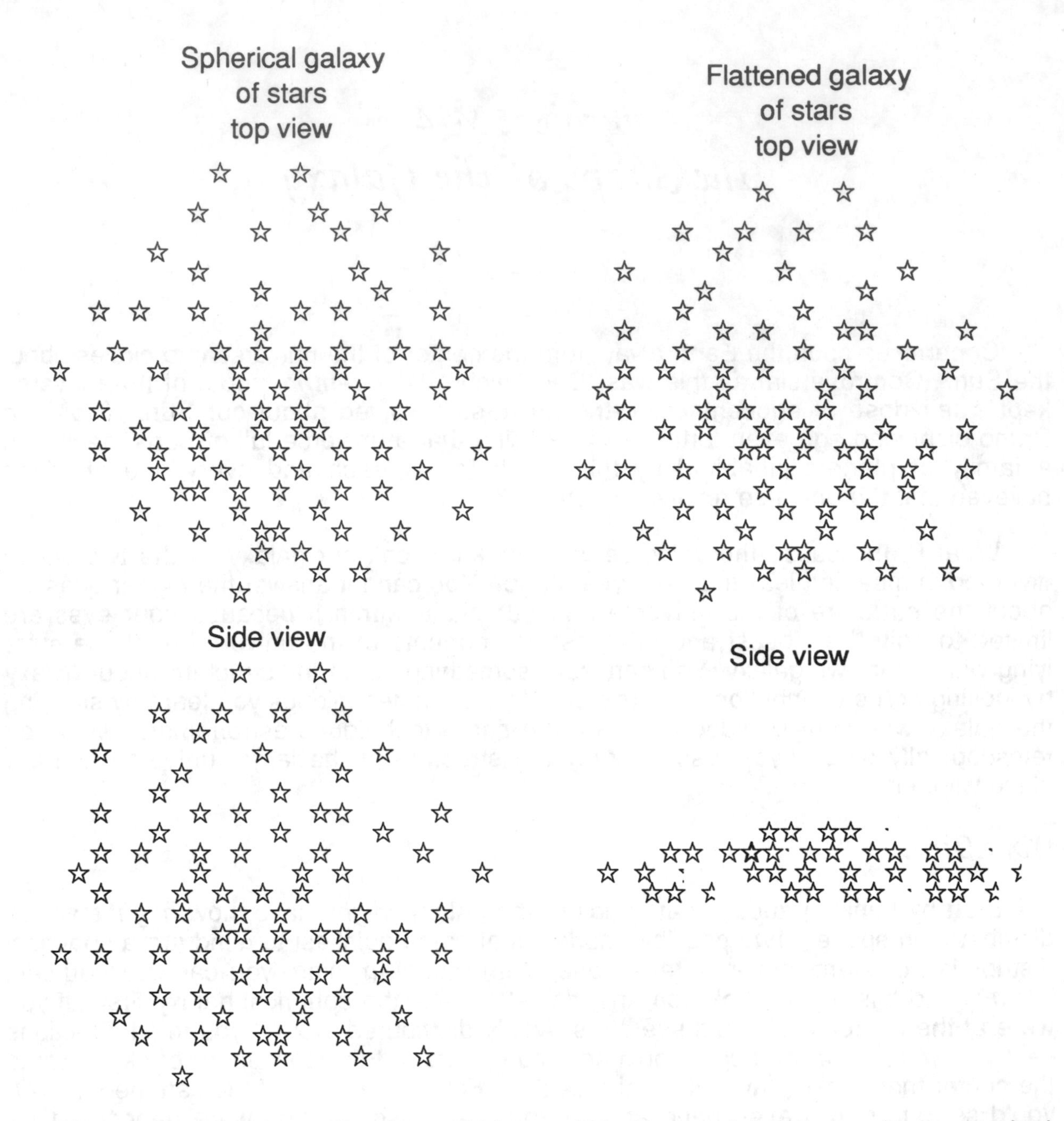

Now that you've mentally observed from within two model galaxies, its time to look outside at a real one; go outside and observe how stars spread across the canvas of night. Yes, there are so many that you can't possibly observe them all! Well said and true. You don't need to sample the whole sky to discover the distribution of stars, just representative parts. For that you need a specialized tool called a *star gauge* — any cardboard tube 1 inch in diameter and about 4 inches long. The star guage limits your field of view to a patch of sky about 15° in diameter. To use it you just look through it at a constellation, count the stars you see, and plot the counts on a map of the sky. Those numbers on the sky map reveal the distribution of stars in different parts of the sky

Star Gauging

1) Table 1 lists prominent constellations evenly spread across the sky. Notice that not all of them are available at the same time of night or from your location on Earth. You may have to work through different times of night and season to complete the table.

Table 1

Constellation Name	RA h	Dec °	Count 1	Count 2	Count 3	Count Average
Pegasus	0	20				
Cassiopeia	0	60				
Andromeda	1	35				
Cetus	2	0				
Aries	2.5	25				
Perseus	3.5	45				
Taurus	4.5	15				
Auriga	5.5	40				
Orion	5.5	0				
Lepus	5.5	-20				
Canis Major	7	-20				
Gemini	7.5	25				
Canis Minor	7.5	5				
Hydra	9.5	-10				
Leo	10.5	20				
Ursa Major	12	55				
Corvus	12.5	-20				
Virgo	13	0				
Bootes	14.5	30				
Libra	15	-15				
Corona Borealis	15.5	30				

Table 1 Con't

Constellation Name	RA h	Dec °	Count 1	Count 2	Count 3	Count Average
Ursa Minor	15.5	75				
Hercules	17	35				
Scorpio	16.5	-25				
Hercules	17	35				
Draco	18	55				
Sagitarius	18.5	-30				
Lyra	19	35				
Aquila	19.5	10				
Cygnus	20	40				
Capricornus	21	-20				
Cepheus	22	65				
Aquarius	23	-10				

2) Pick a bright star in each constellation, look through your star gauge at the constellation and count the number of stars you see. Record that number in Table 1.

3) Repeat step 2 for two other bright stars in the constellation and then average your three values for the constellation.

4) Repeat steps 2 and 3 for all the constellations in Table 1.

5) Write the number of stars for each constellation in the following figure to see the distribution of stars in different directions of the sky. Remember, since this is a flat map, you can expect some distortions.

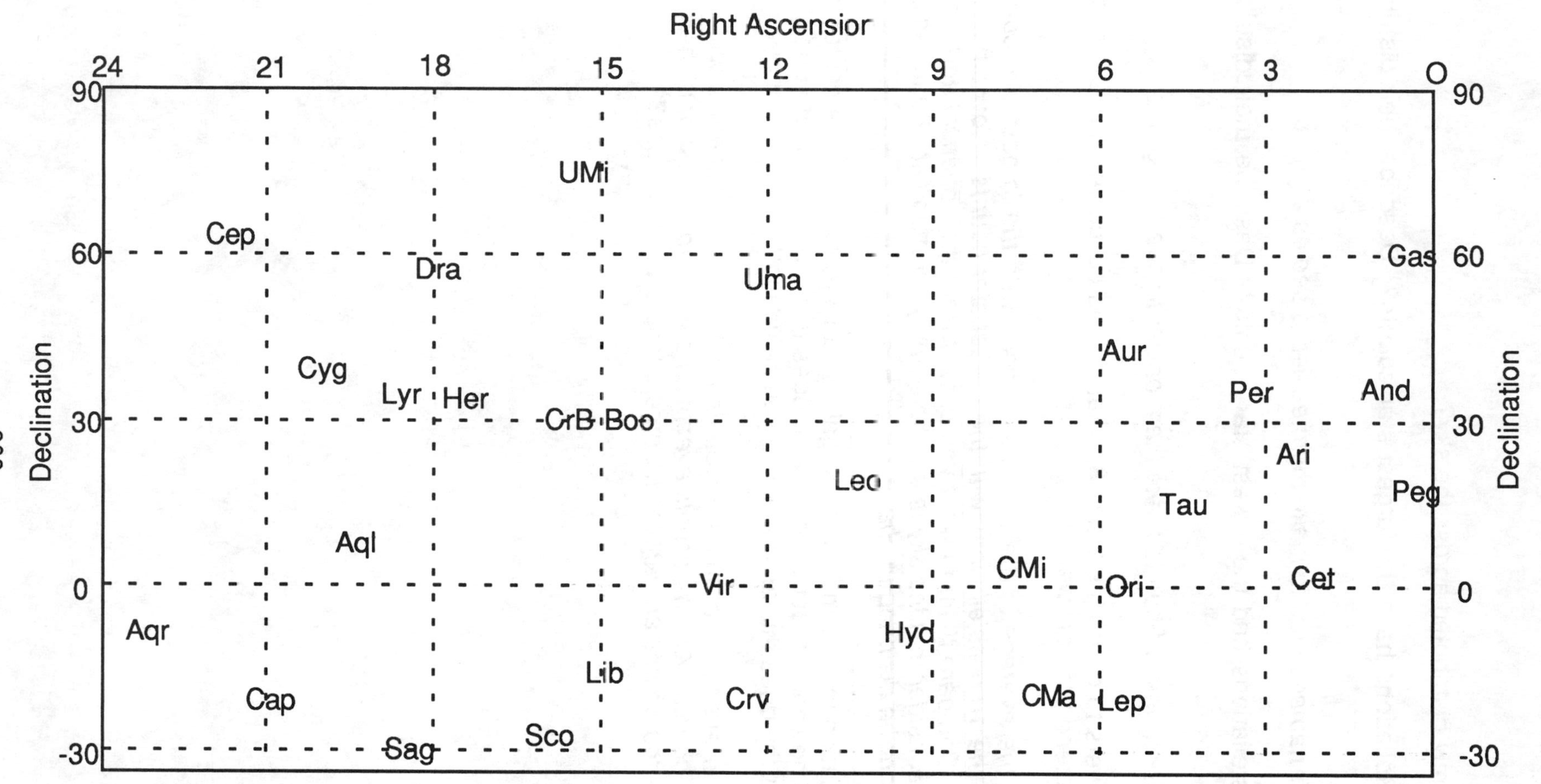
Right Ascensior
24
21
18
15
12
9
6
3
O
90
60
30
0
-30
Declination
Declination
UMi
Cep
Dra
Uma
Gas
Aur
Cyg
Lyr
Her
CrB
Boo
Per
And
Ari
Leo
Tau
Peg
Aql
CMi
Vir
Ori
Cet
Aqr
Hyd
Lib
Cap
Crv
CMa
Lep
Sco
Sag

Questions

1) Are stars uniformly distributed across the sky?

2) Which 10 constellations have the largest star counts? Describe their distribution across the sky.

   Hint: Circle their names in red to make them easier to see as a group.

3) Which 10 constellations had the lowest star counts? Describe their distribution across the sky.

   Hint: Circle their names in Bbue to make them easier to see as a group.

4) From your answers to questions 2 and 3, what do you conclude as to the shape of our galaxy? Defend your answer.

   Hint: Compare the average of the 10 constellations with the largest star counts to the average of the 10 constellations with the lowest star counts. How different are the averages? Compare them to what you expect from the spherical and flattened models. Also consider that the sky is a sphere and your map flat. Features on the map will therefore be distorted in shape.

5) What is our location in your model of the galaxy (are we close to the center, out at the edge or somewhere in between)? What does this indicate about your place in the universe? Why does this answer strike you as wrong?

   Hint: Look at the evenness of the star counts for the 10 constellations which have the largest star counts. Are the counts even all the way around us, or is there some direction where the counts are larger and others where they are less?

Project A-5
Intrinsic Variable Stars

Not all phases of stellar evolution are slow. Supernovae provide an awe inspiring example of titanic evolutionary changes in a matter of hours, but supernovae are few and far between. There are other stars — some near the ends of their lives, others not yet born onto the main sequence — that change their luminosities, temperatures and radii regularly over the course of hours or days. They reveal the phases in a star's life where the delicate balance between gravity and pressure is unstable.

Table 1 lists variable stars of several different types, many of the names for the types taken from the first recognized example. Thus we have the classical cepheids (Cδ) named after δ Cephei, W Virginis type cepheids (CW), RR Lyrae type variables, (RRa, RRb), Dwarf Cepheids (RRs — originally thought to be extreme cases of RR Lyrae type, but now recognized as distinct), RV Tauri types (RVa, RVb), and long-period semi-regular cool giants (SR). All are more luminous than the Sun and all reflect late stages in a star's evolution. Some types vary in hours, others in months. Some types have regular repeatable cycles of variation, while others are erratic. Each type lives in a different region of the H-R diagram.

Table 1

Name	R.A.			Dec			Peak Visual Magnitude	Period (days)	Variable Type
	h	m	s	°	'	"			
TU Cas	0	26	19	51	16	49	7.1	2.1393	CW
ST Tau	5	45	3	13	34	12	7.8	4.0342	CW
BQ Ori	5	57	07	22	50	21	6.7	110.	SR
W Gem	6	34	57	15	19	50	6.5	7.9141	Cδ
U Mon	7	30	47	-9	46	38	6.0	92.26	RVb
VZ Cnc	8	40	52	9	49	27	7.5	0.1784	RRs
V702 Sco	17	42	16	-32	31	24	7.6	0.1152	RRa
AC Her	18	30	16	21	52	00	7.4	75.4619	RVa
U Sgr	18	31	53	-19	07	30	6.4	6.7452	Cδ
RR Lyr	19	25	28	42	47	04	6.9	0.5667	RRa
AF Cyg	19	30	13	46	08	53	7.4	94	SR

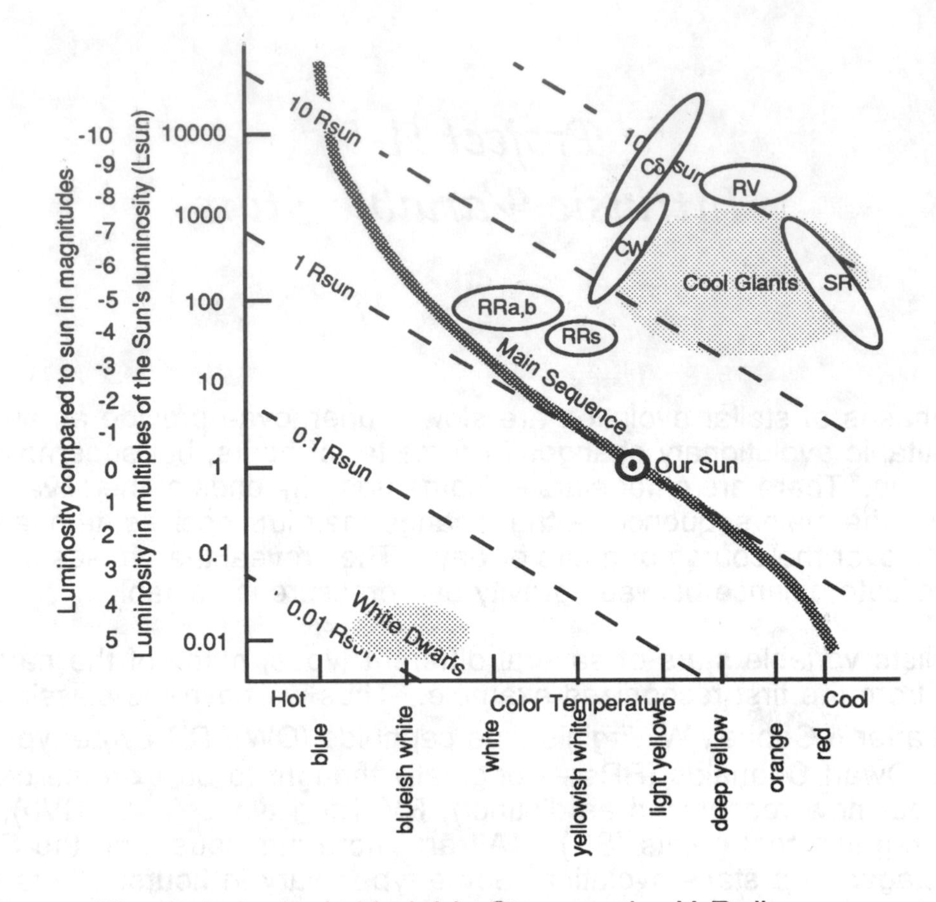

Figure 1: Intrinsic Variable Stars on the H-R diagram

In all these stars the struggle between pressure and gravity is made quite visible as their cycles of pulsation reveal themselves in cycles of luminosity, temperature and size — as cycles of visual magnitude and color. As you watch their brightnesses change, you are seeing them move on the H-R diagram.

Observe their visual magnitudes from hour to hour or day to day to trace their motion on the H-R diagram. Use the same visual photometry technique as for Project 4-2. Finding charts, covering all stars 8th magnitude or brighter in a 10° diameter circle about each variable star, are provided for locating the variable stars. Comparison charts, covering a 2° or 3° diameter circle about each variable star, give standard stars for comparison through the eyepiece of your telescope. Note the lack of decimal points to avoid confusion with faint stars in the comparison charts; a 10.1 magnitude star will be labeled 101, a 9.4 magnitude star 94. Use Table 2 to record your observations.

Not only will the variables change by different amounts during their pulsation, but the shapes of their light curves (magnitude plotted versus time) differ. Some stars brighten rapidly and fade slowly, others brighten slowly and fade rapidly, etc.

Tricks of the Trade

1) *Purge your memory each time you measure a star's magnitude. Don't pollute tonight's observation with memories of how bright it looked relative to the comparison standards last time.*

2) *Don't zero in on your star too fast by visually picking one that at first glance seems to match .*

3) *The last two comparision standards that bracket each unknown should not differ by more than 0.5 magnitudes.*

4) *Look for small patterns of stars (triangles, squares, pentagons, etc...) to find your way around .*

5) *Try using higher power eyepieces on the fainter stars.*

Questions

1) How many magnitudes did each variable change by? Which was quicker: the change from brightest to faintest, or faintest to brightest?

2) Is there any relationship between how long the variables took to vary (period) and the change in visual magnitude over the cycle?

Hint: Plot magnitude swing versus period for all stars you've observed.

3) Estimate the period of each variable from your observations and compare your results to the periods listed in Table 1.

Table 2: Variable Observations

Obs Number	Date	Time	RA h	m	Dec °	'	Alt °	Azi °	Visual Mag

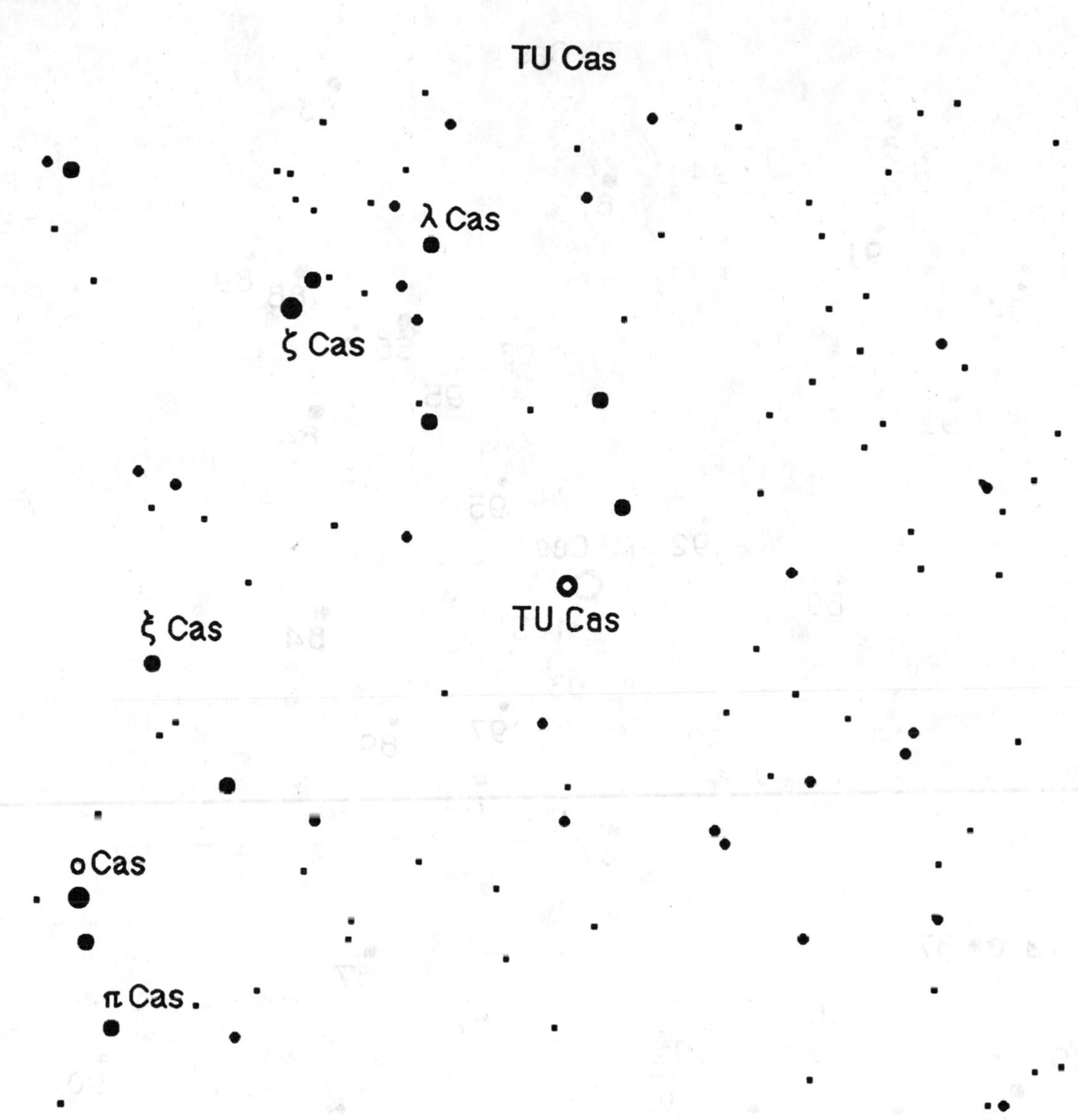
TU Cas
λ Cas
ζ Cas
ξ Cas
TU Cas
o Cas
π Cas
10° x 10°

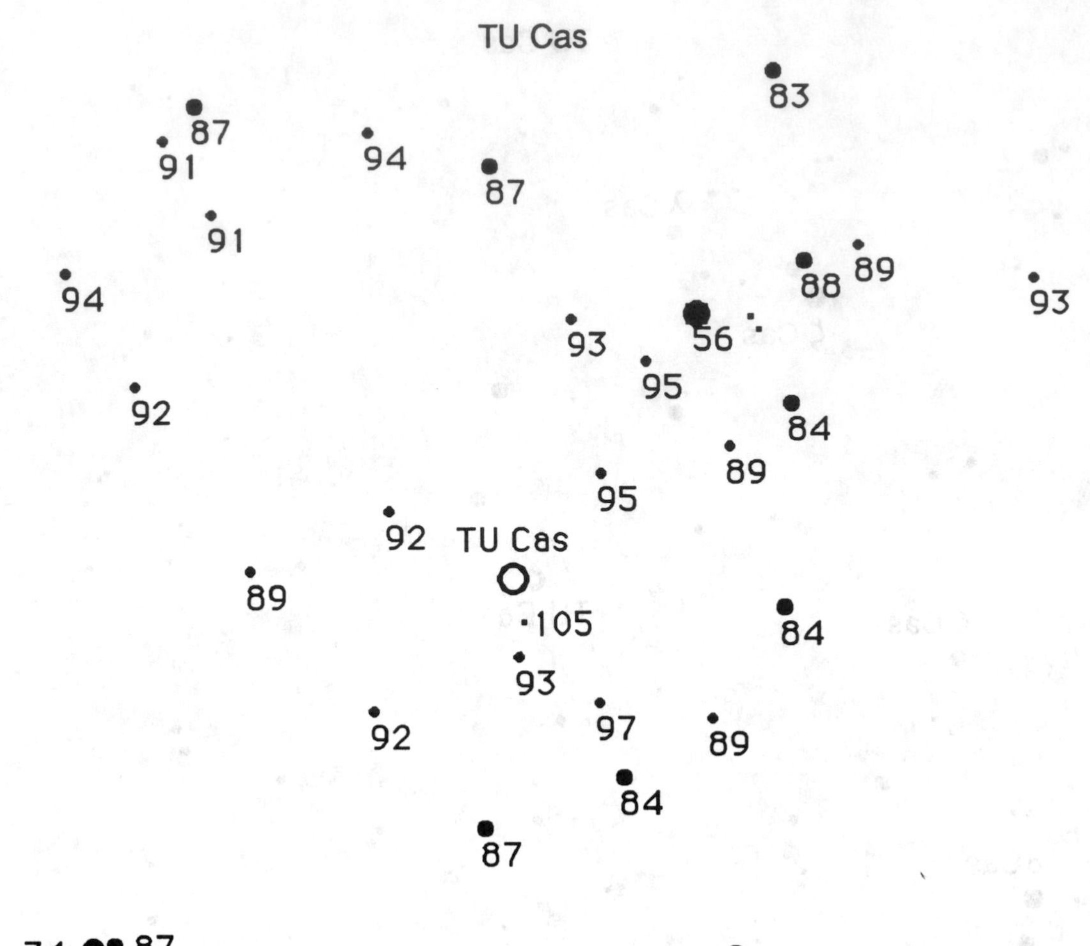
TU Cas
83
87
91
94
87
91
89
88
94
93
56
93
95
92
84
89
95
92
TU Cas
89
105
84
93
92
97
89
84
87

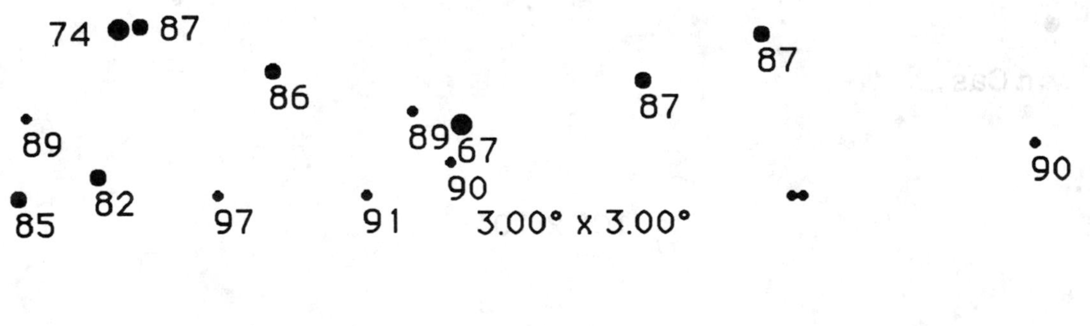
74
87
87
86
87
89
89
67
90
90
85
82
97
91
3.00° x 3.00°

ST Tau

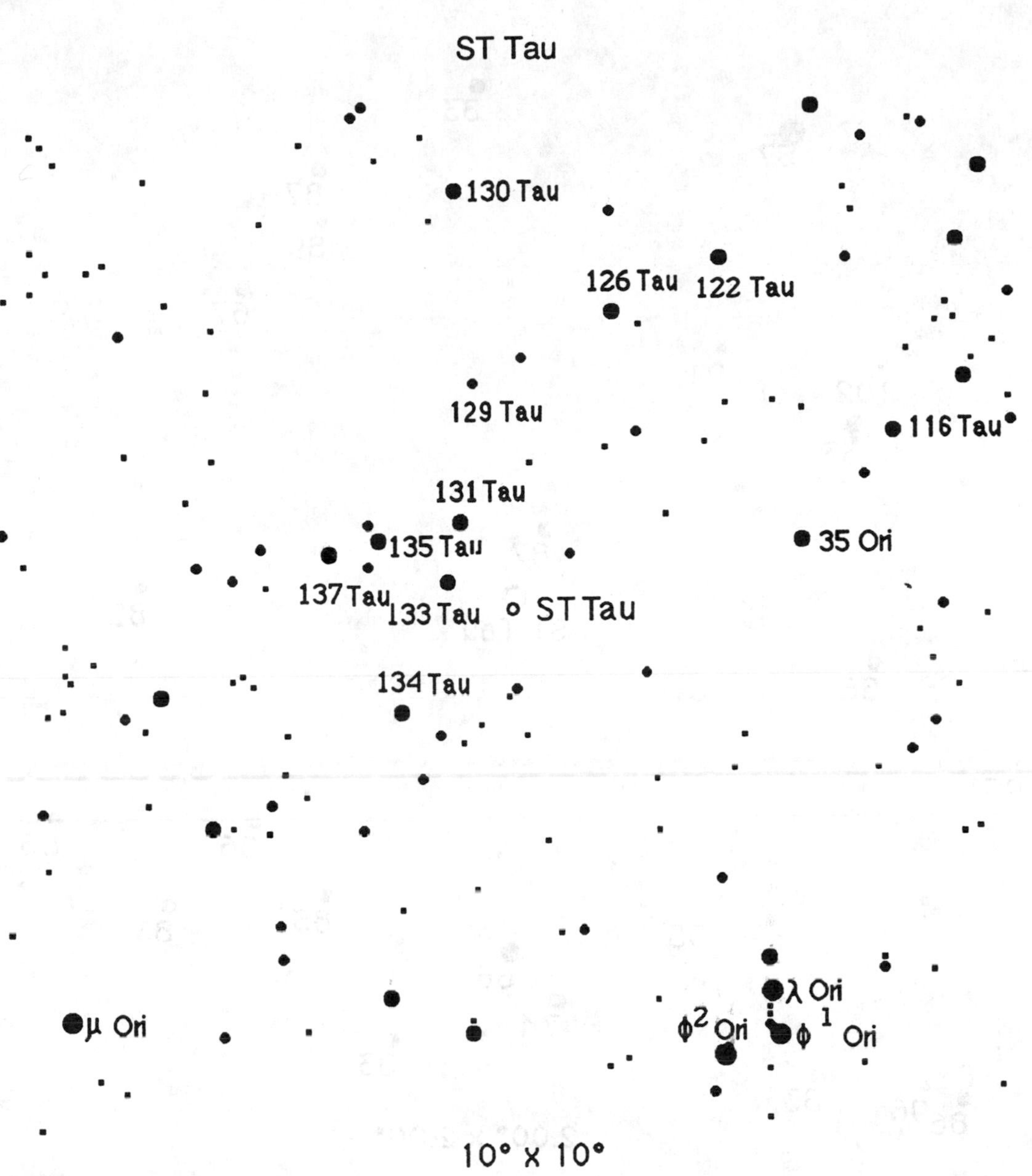
130 Tau
126 Tau
122 Tau
129 Tau
116 Tau
131 Tau
135 Tau
35 Ori
137 Tau
133 Tau
ST Tau
134 Tau
λ Ori
μ Ori
φ2 Ori
φ 1 Ori

10° x 10°

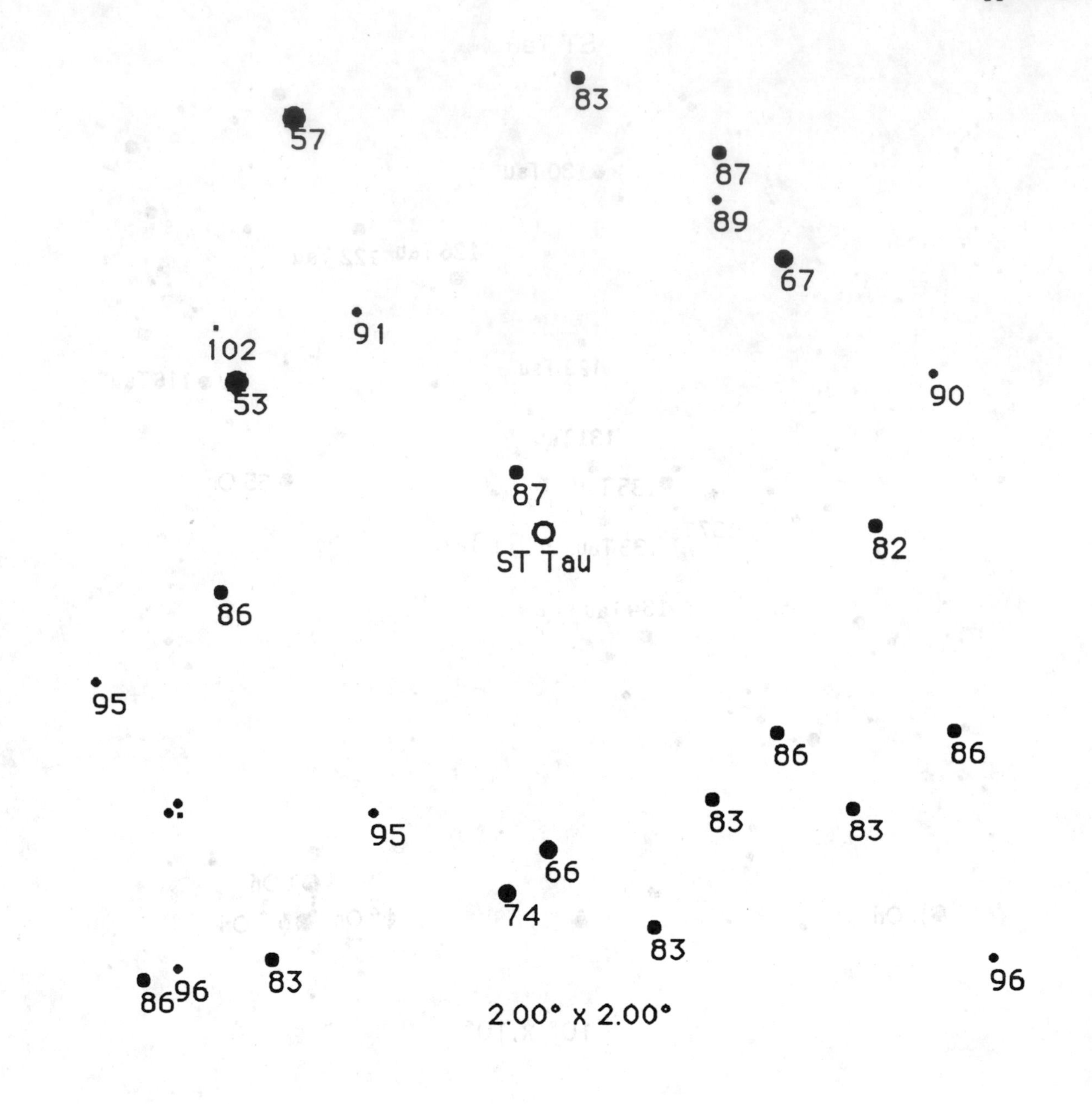
83
57
87
89
67
91
102
53
90
87
ST Tau
82
86
95
86
86
95
83
83
66
74
83
83
96
86
96
2.00° x 2.00°

BQ Ori

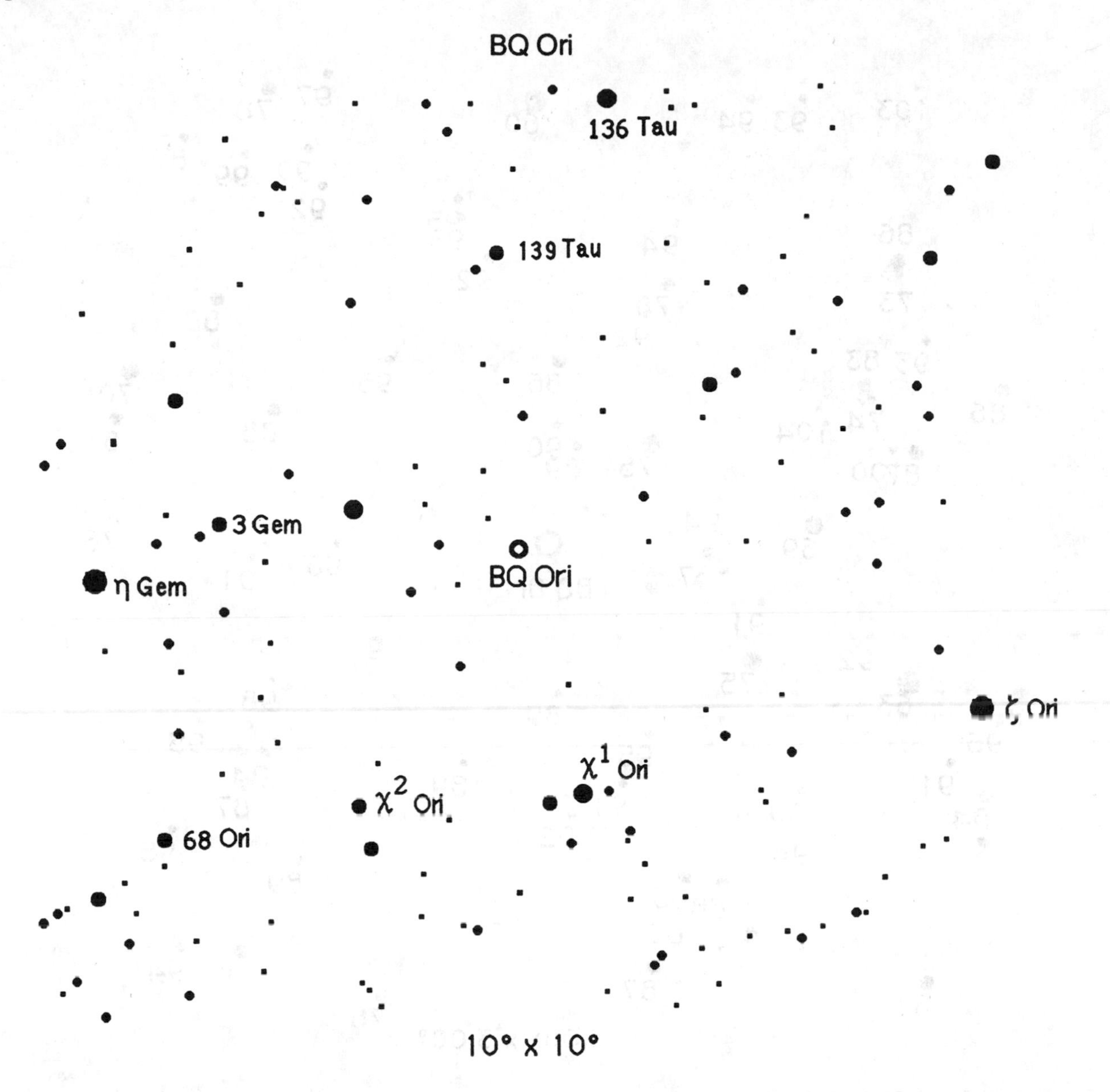

10° x 10°

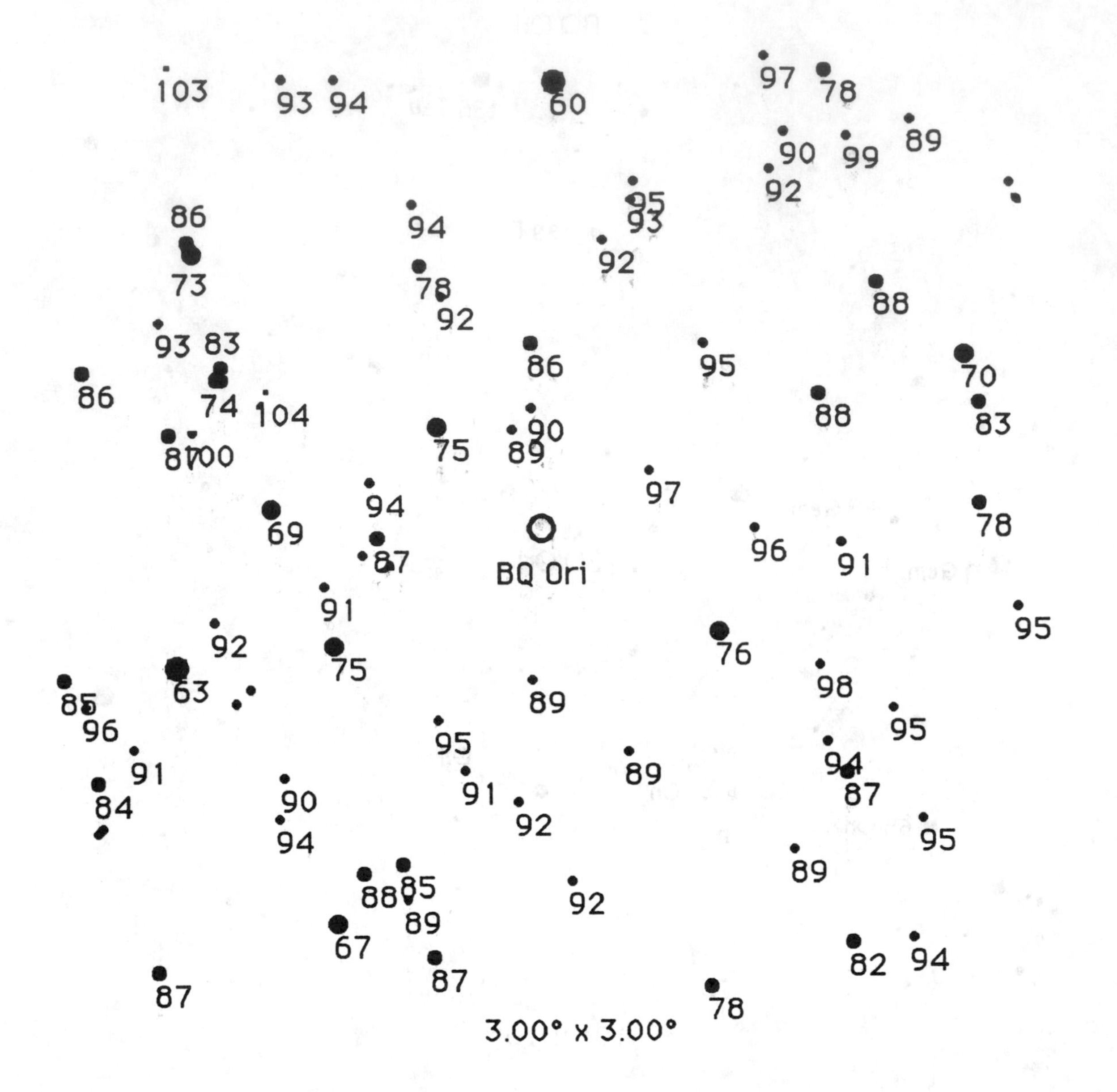
BQ Ori
3.00° x 3.00°

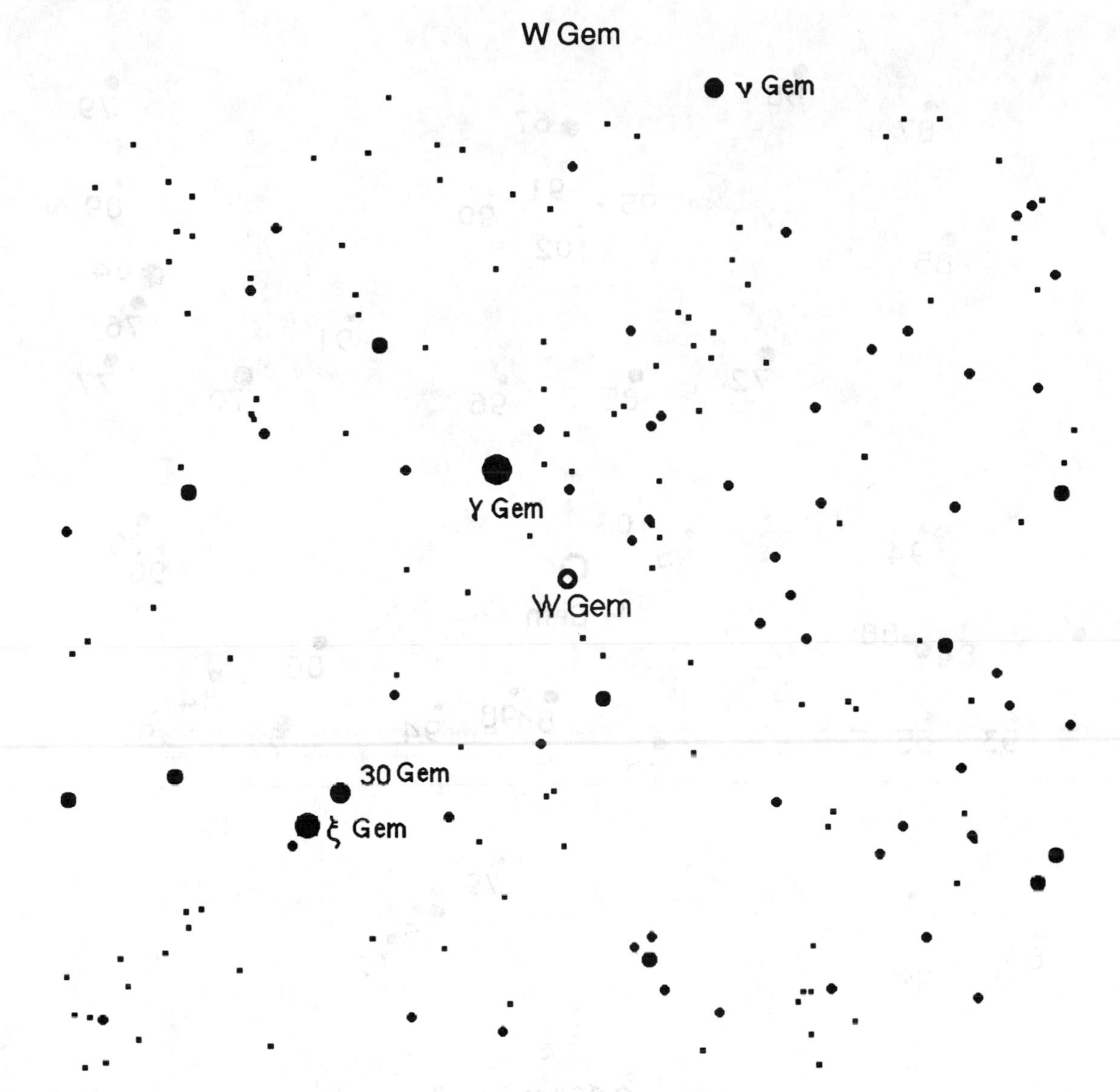
W Gem
ν Gem
γ Gem
W Gem
30 Gem
ξ Gem
10° x 10°

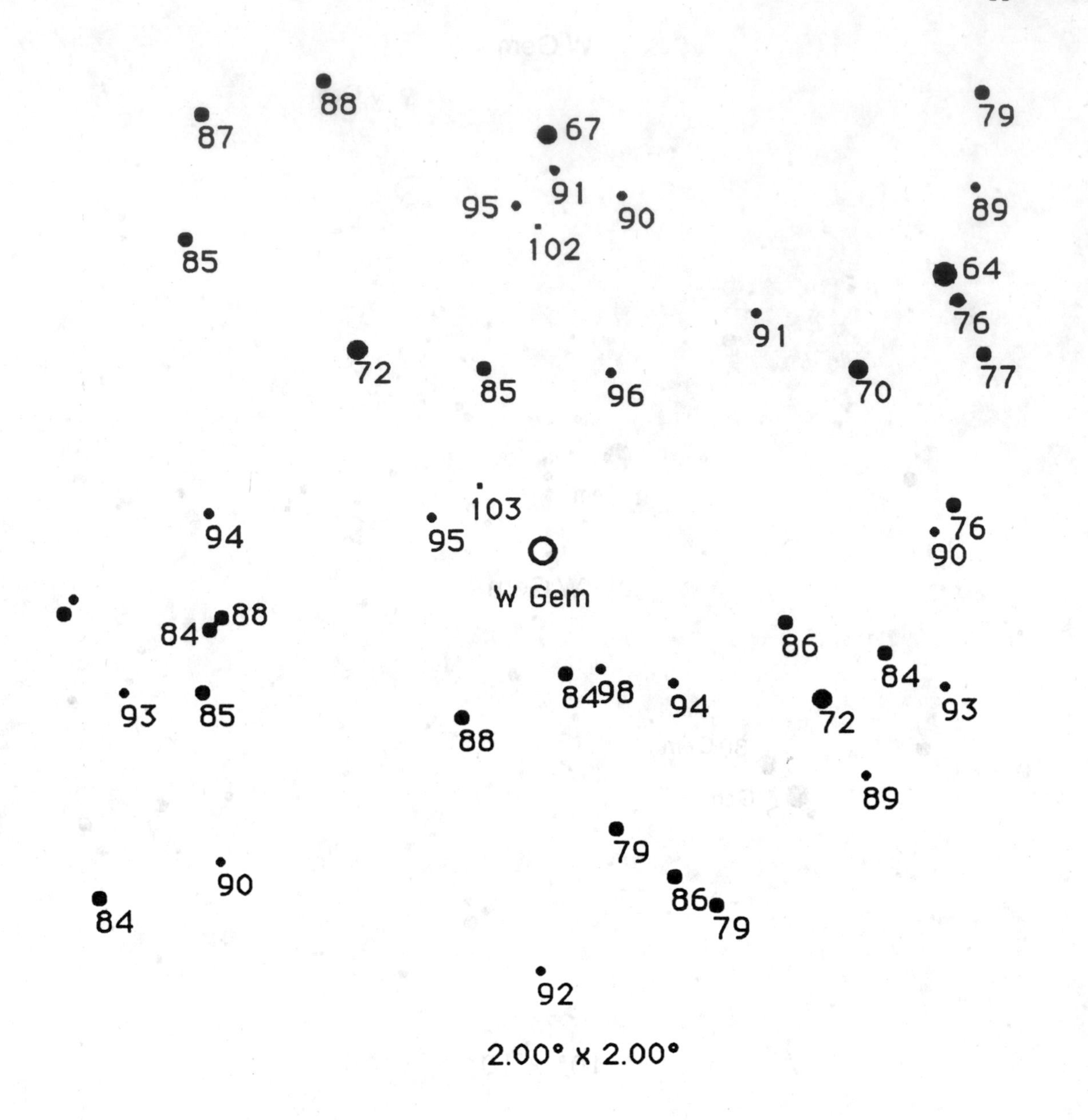
88
87
67
79
91
95
90
89
102
85
64
76
91
72
85
96
70
77
103
76
94
95
90
W Gem
88
84
86
84
84
98
94
93
93
85
72
88
89
79
90
86
84
79
92
2.00° x 2.00°

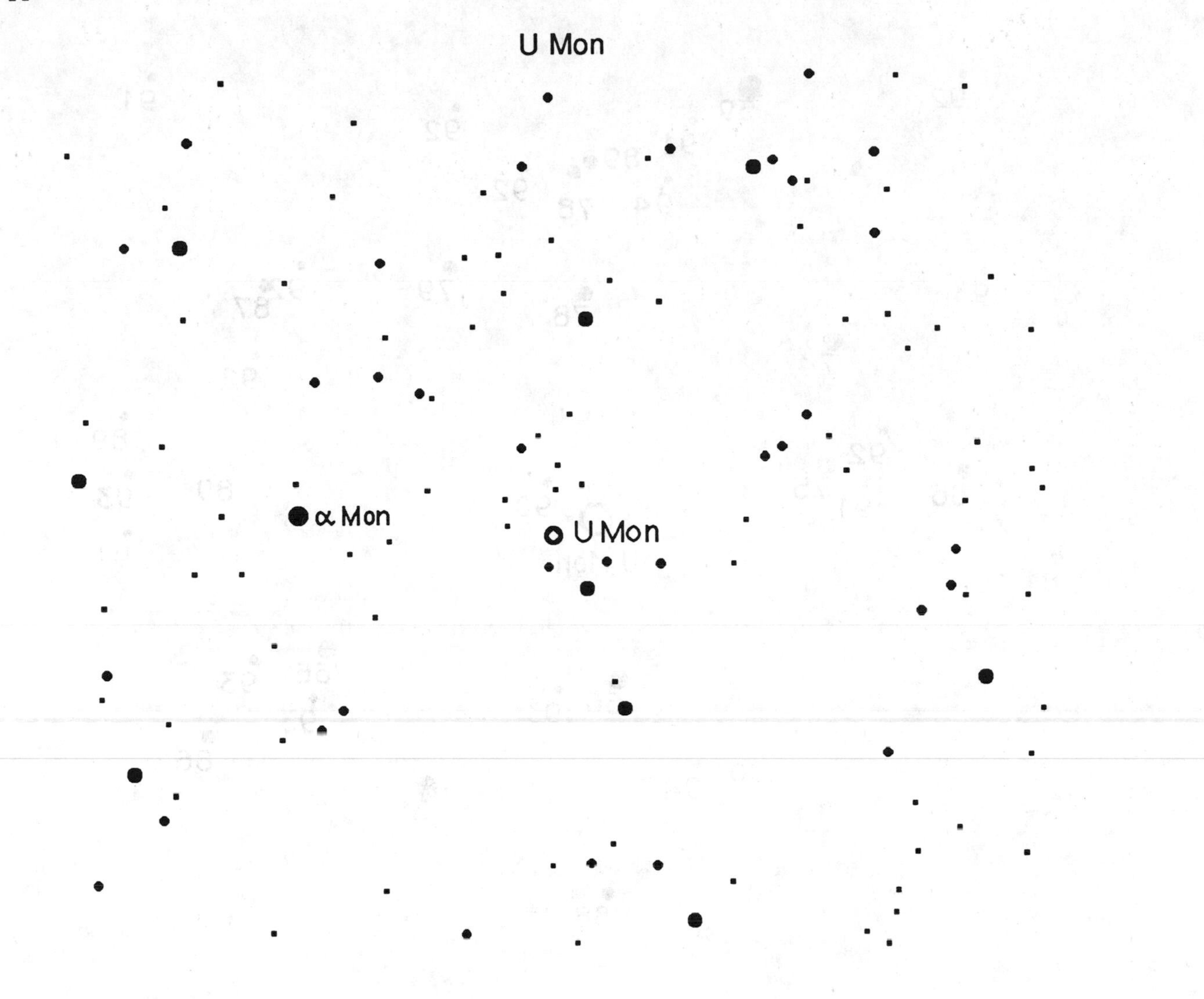

10° x 10°

92
59
91
92
91
89
78
92
94
93
79
92
87
78
74
92
89
92
86
75
91
90
89
93
U Mon
81
91
66
93
66
93
93
86
90
94
58
93
89
84
88

2.00° x 2.00°

VZ Cnc

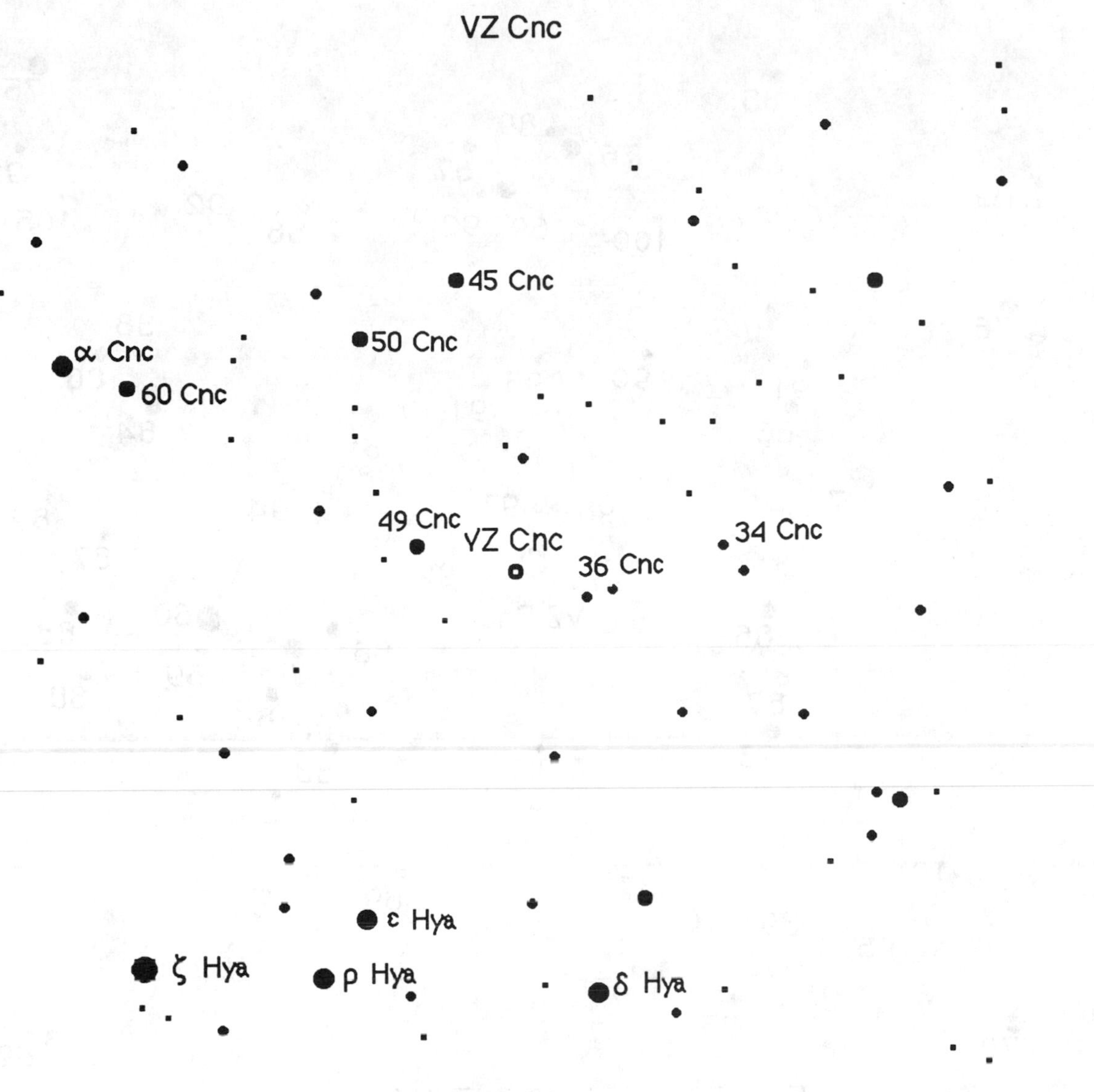
45 Cnc
50 Cnc
α Cnc
60 Cnc
49 Cnc
VZ Cnc
36 Cnc
34 Cnc
ε Hya
ζ Hya
ρ Hya
δ Hya

10° x 10°

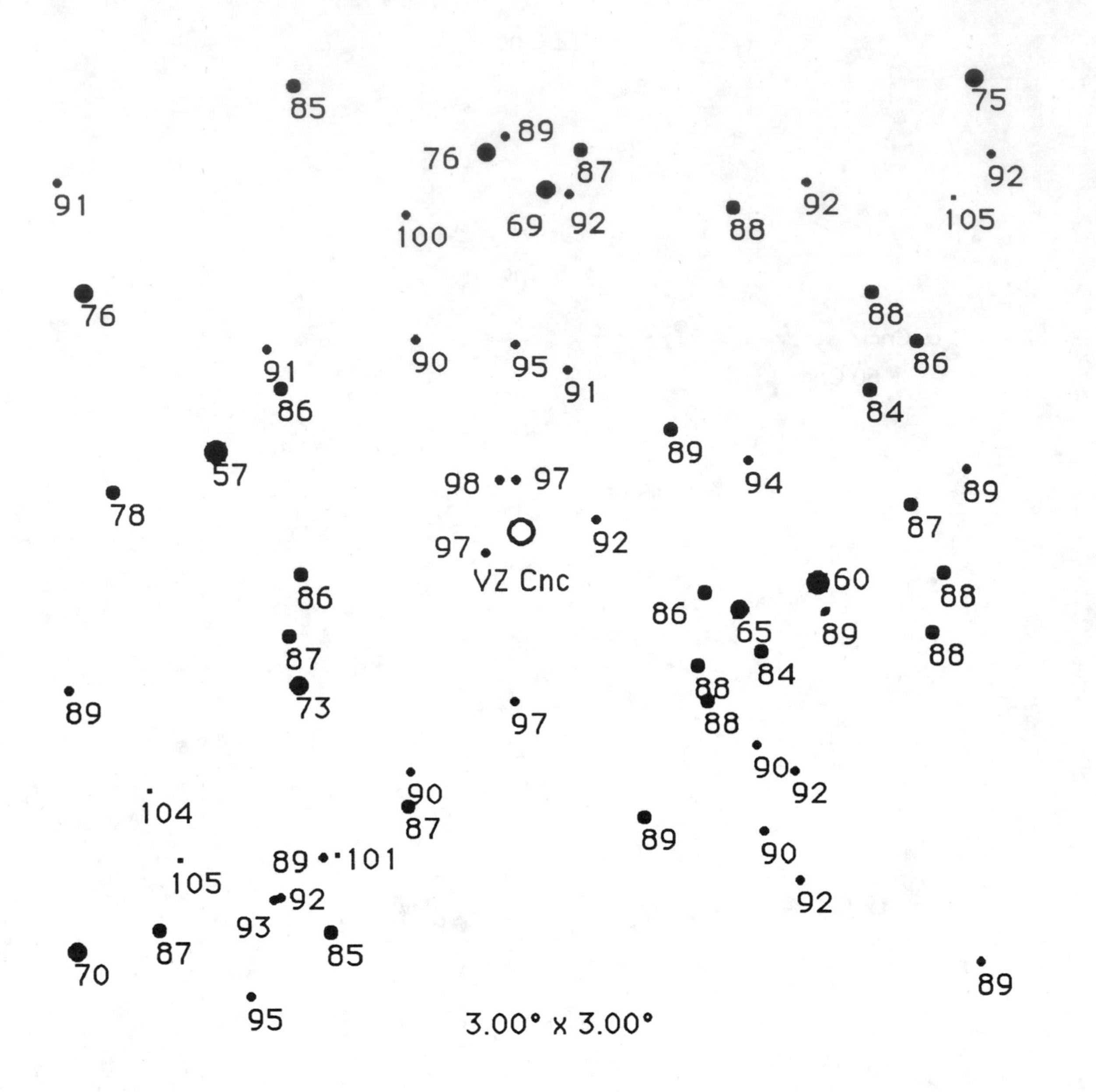
85
75
89
76
87
92
91
69
92
88
92
105
100
88
76
86
91
90
95
91
86
84
57
89
94
89
78
98
97
87
97
92
VZ Cnc
60
86
88
86
65
89
87
84
88
73
88
89
97
88
90
92
90
104
87
89
89
101
90
105
92
92
93
87
85
70
89
95
3.00° x 3.00°

V702 Sco

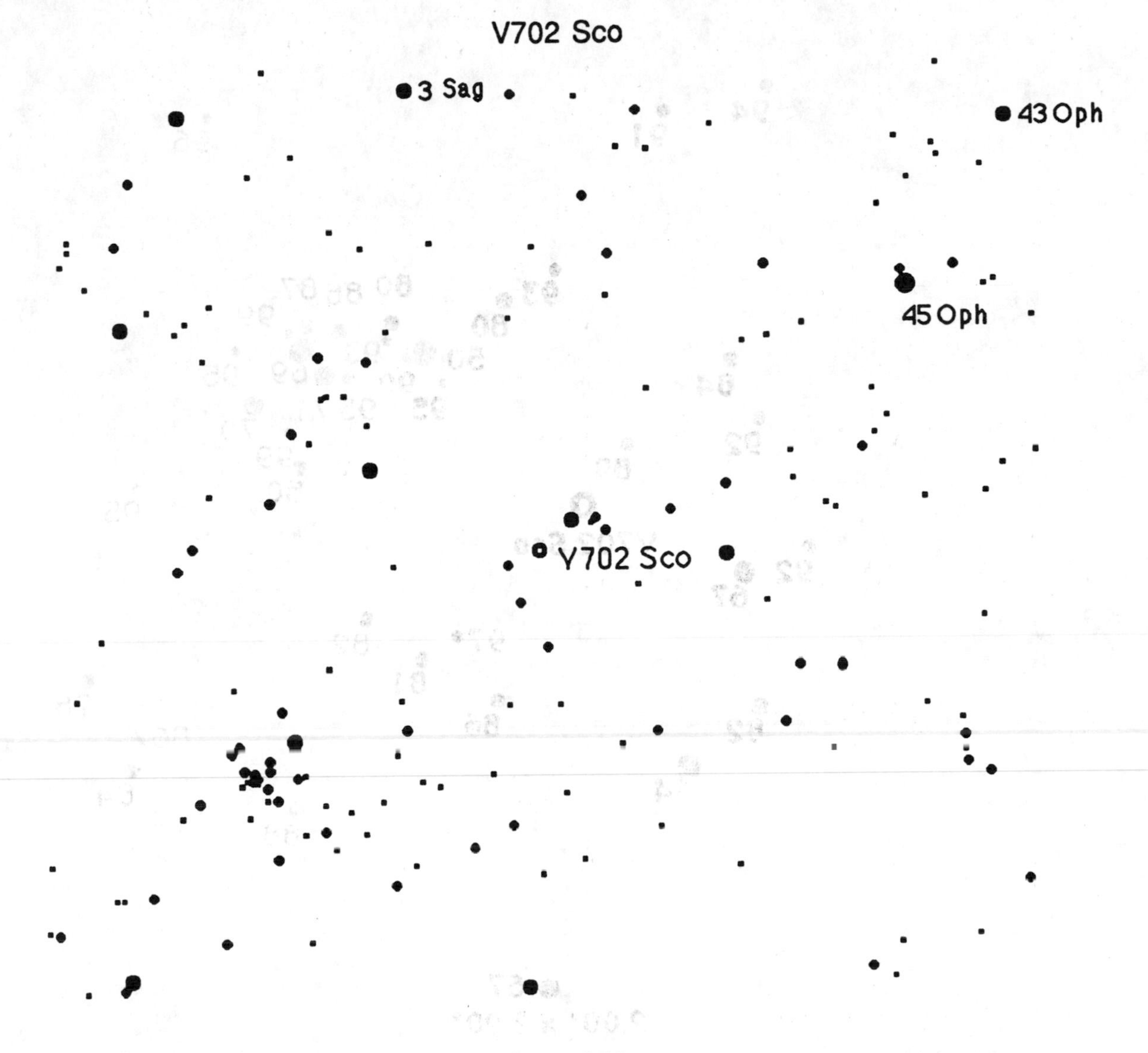

10° x 10°

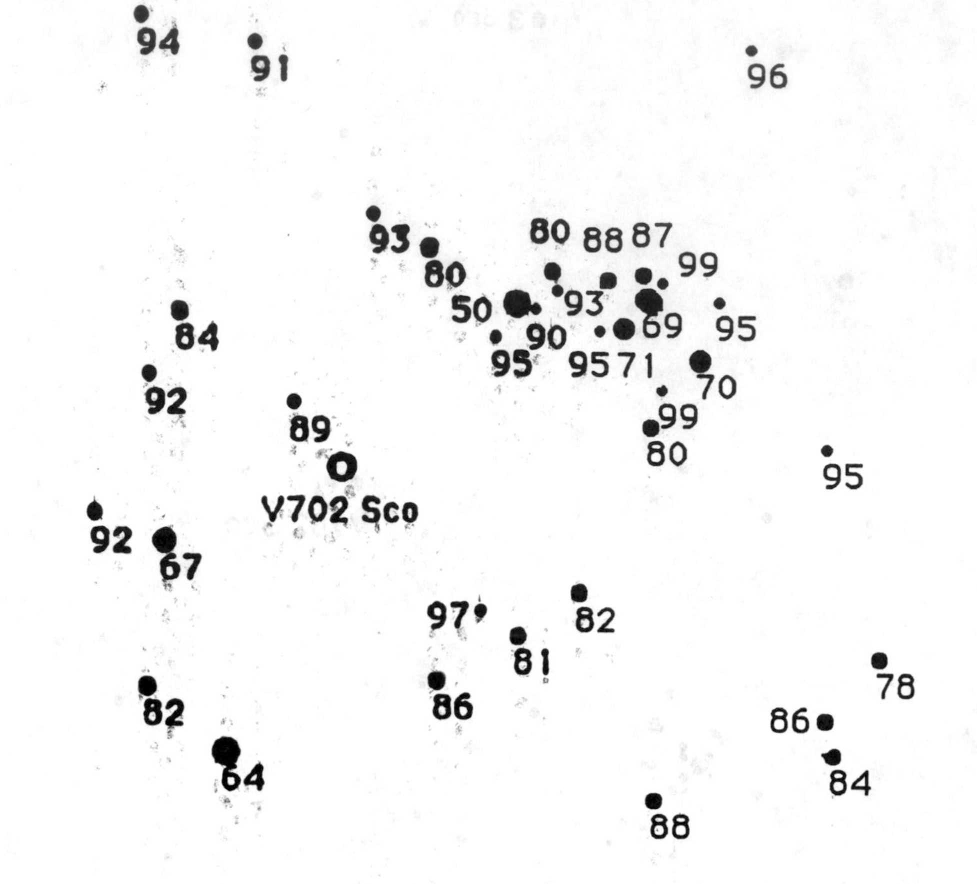
94
91
96
93
80
80
88
87
99
50
93
84
69
95
90
95
95
71
92
70
89
99
80
95
V702 Sco
92
67
97
82
81
78
82
86
86
64
84
88
67
2.00° x 2.00°

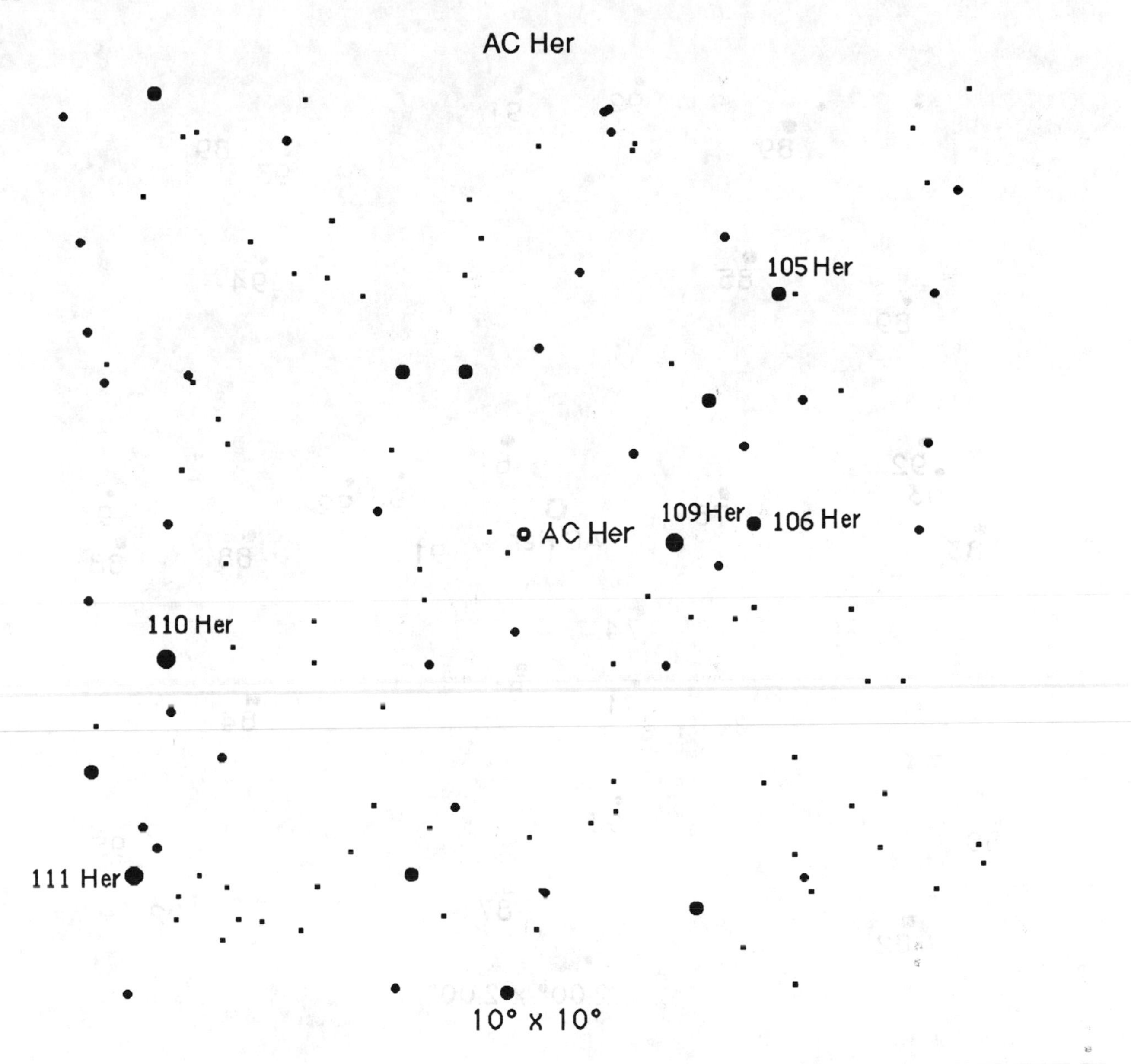
AC Her
105 Her
109 Her
106 Her
AC Her
110 Her
111 Her
10° x 10°

92
91
89
89
93
85
94
89
92
81
93
93
92
78
93
AC Her
91
83
88
88
74
81
91
84
93
89
95
87
92
82
2.00° x 2.00°

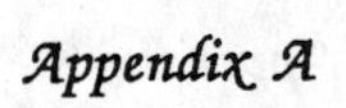

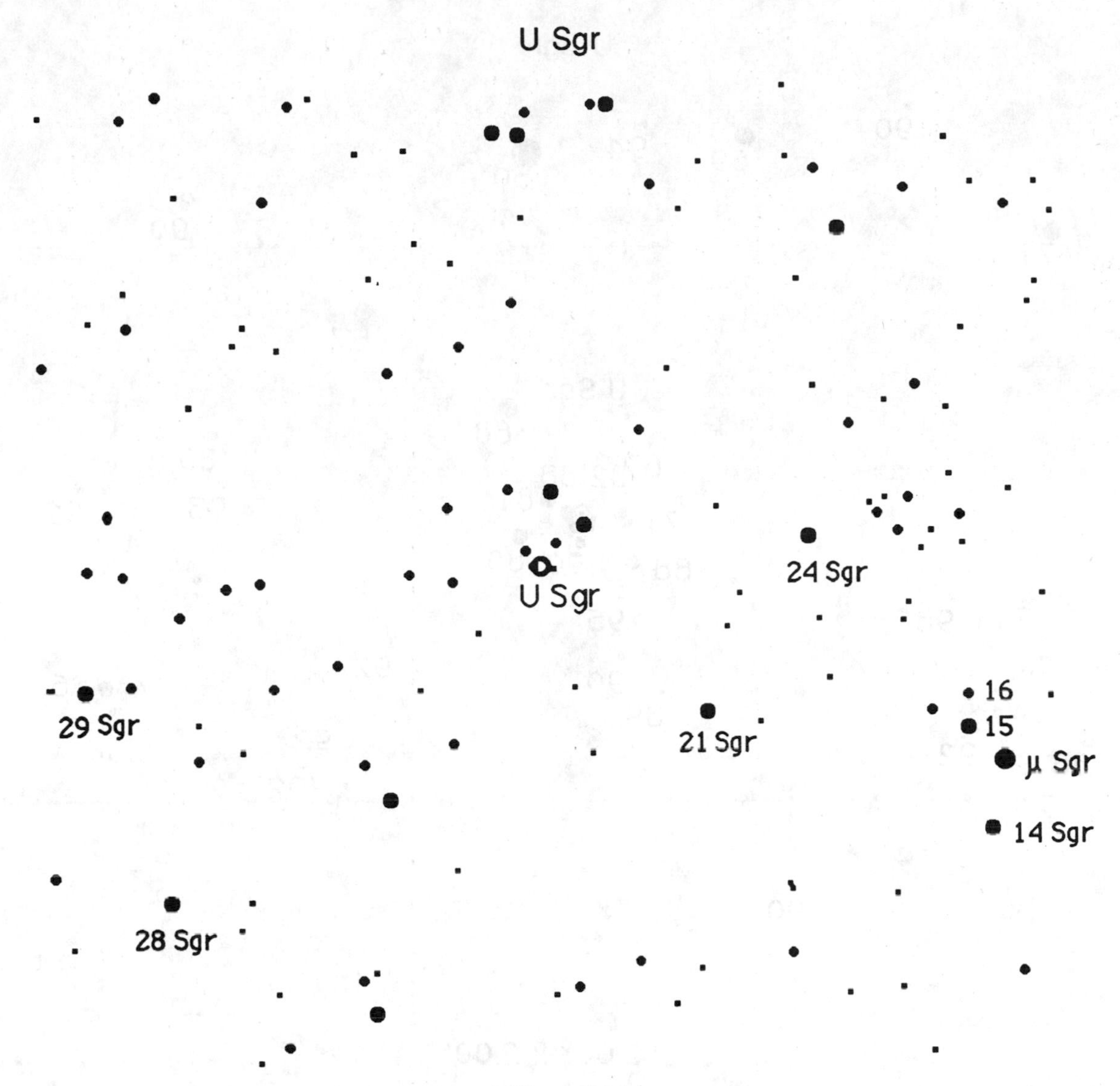
U Sgr
U Sgr
24 Sgr
29 Sgr
21 Sgr
16
15
μ Sgr
14 Sgr
28 Sgr
10° x 10°

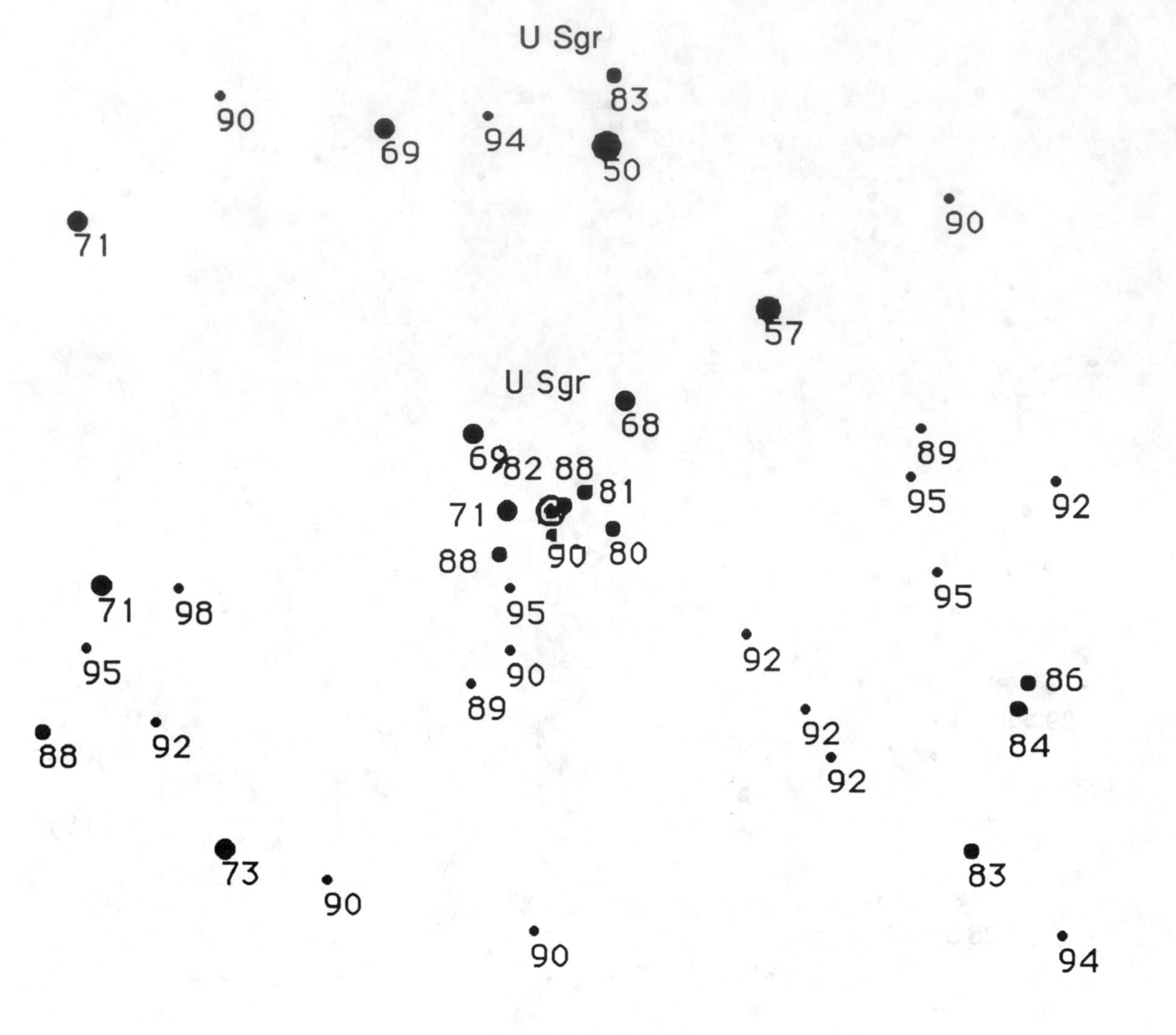
U Sgr
83
90
69
94
50
90
71
57
U Sgr
68
89
69
82
88
81
95
92
71
80
88
90
95
71
98
95
92
90
86
89
92
84
95
88
92
92
73
83
90
90
94
2.00° x 2.00°

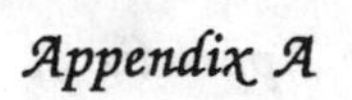

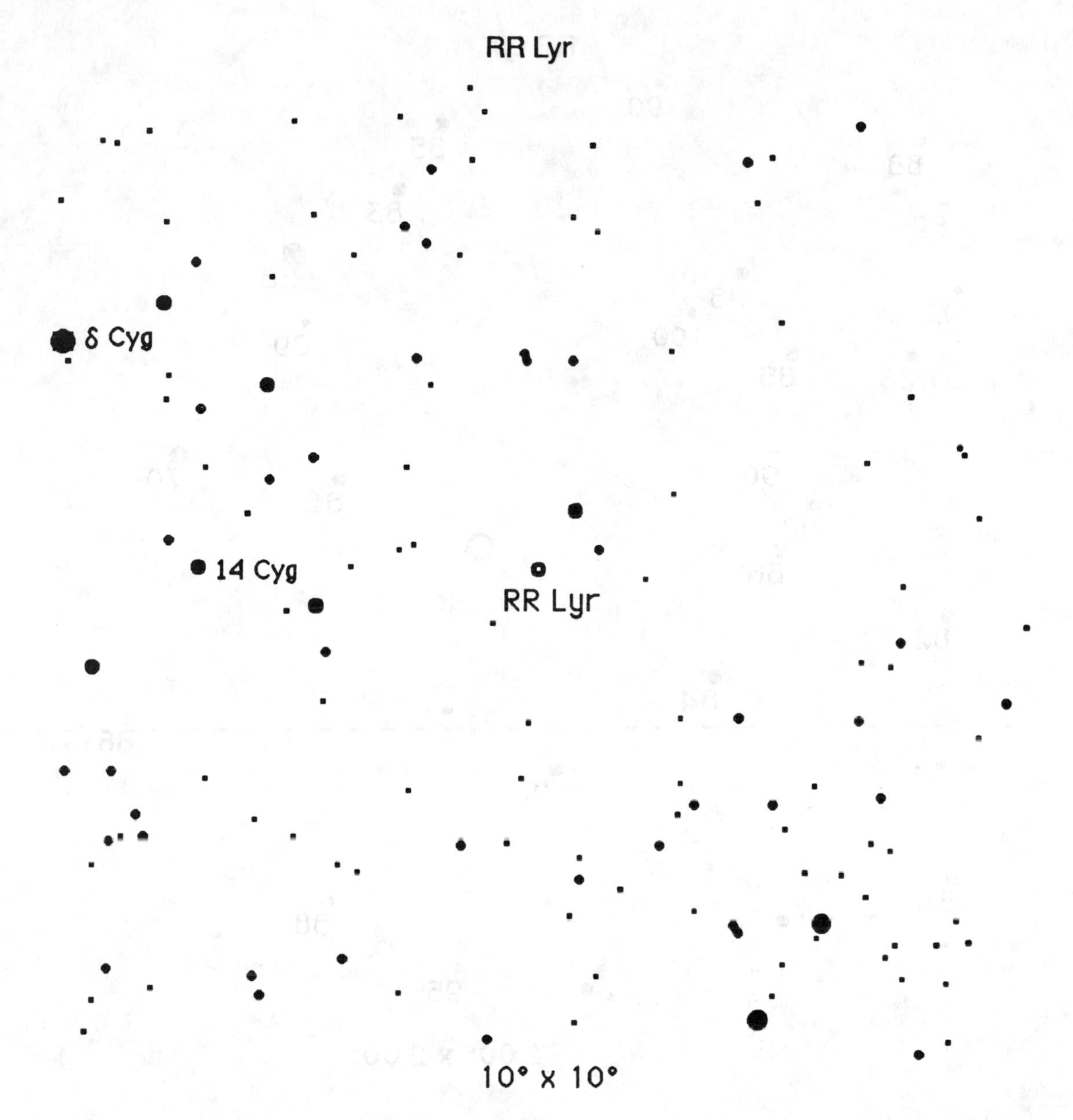
RR Lyr
δ Cyg
14 Cyg
RR Lyr
10° x 10°

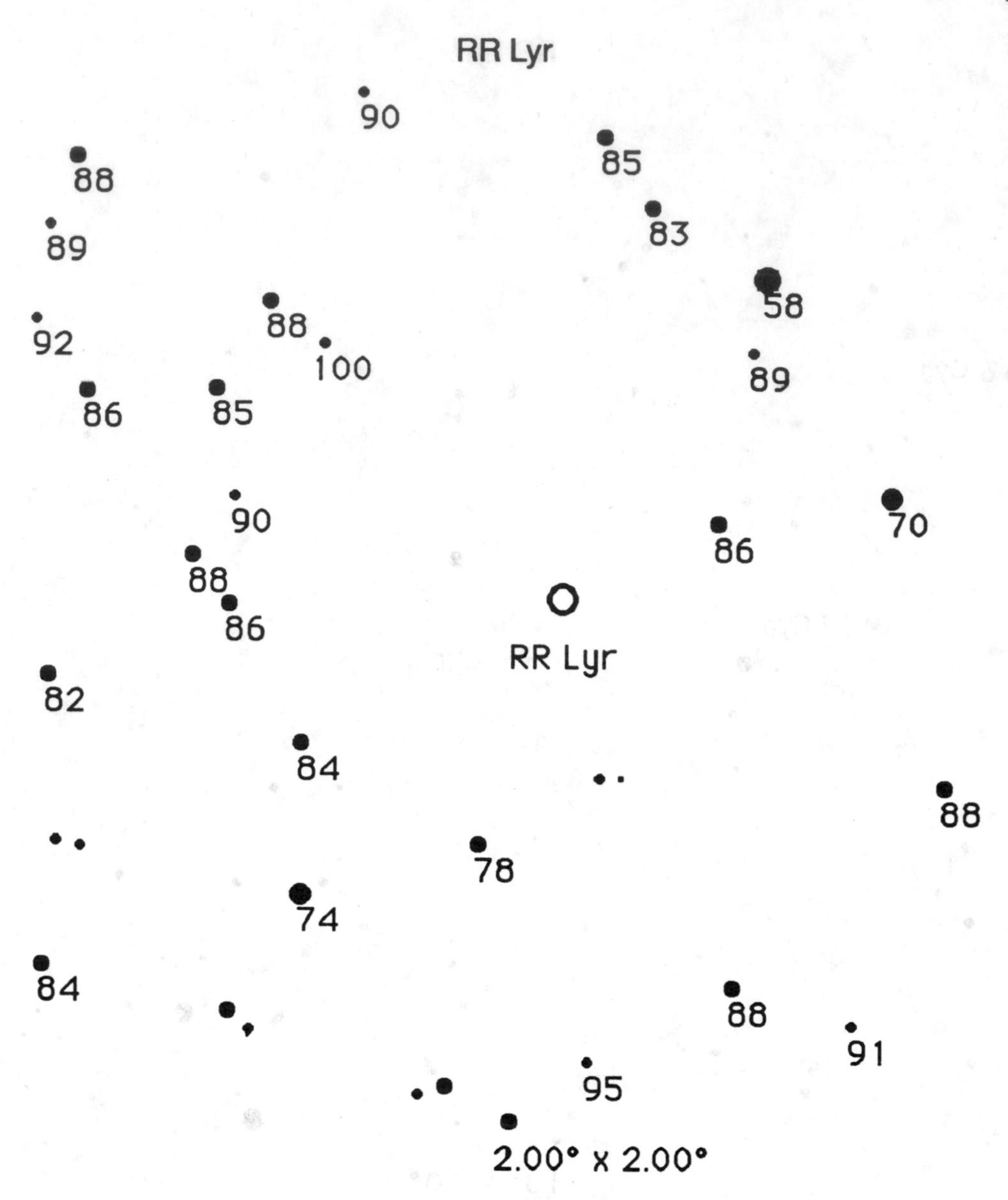
RR Lyr
90
88
85
83
89
58
88
92
100
89
86
85
90
70
86
88
86
RR Lyr
82
84
88
78
74
84
88
91
95
2.00° x 2.00°

AF Cyg

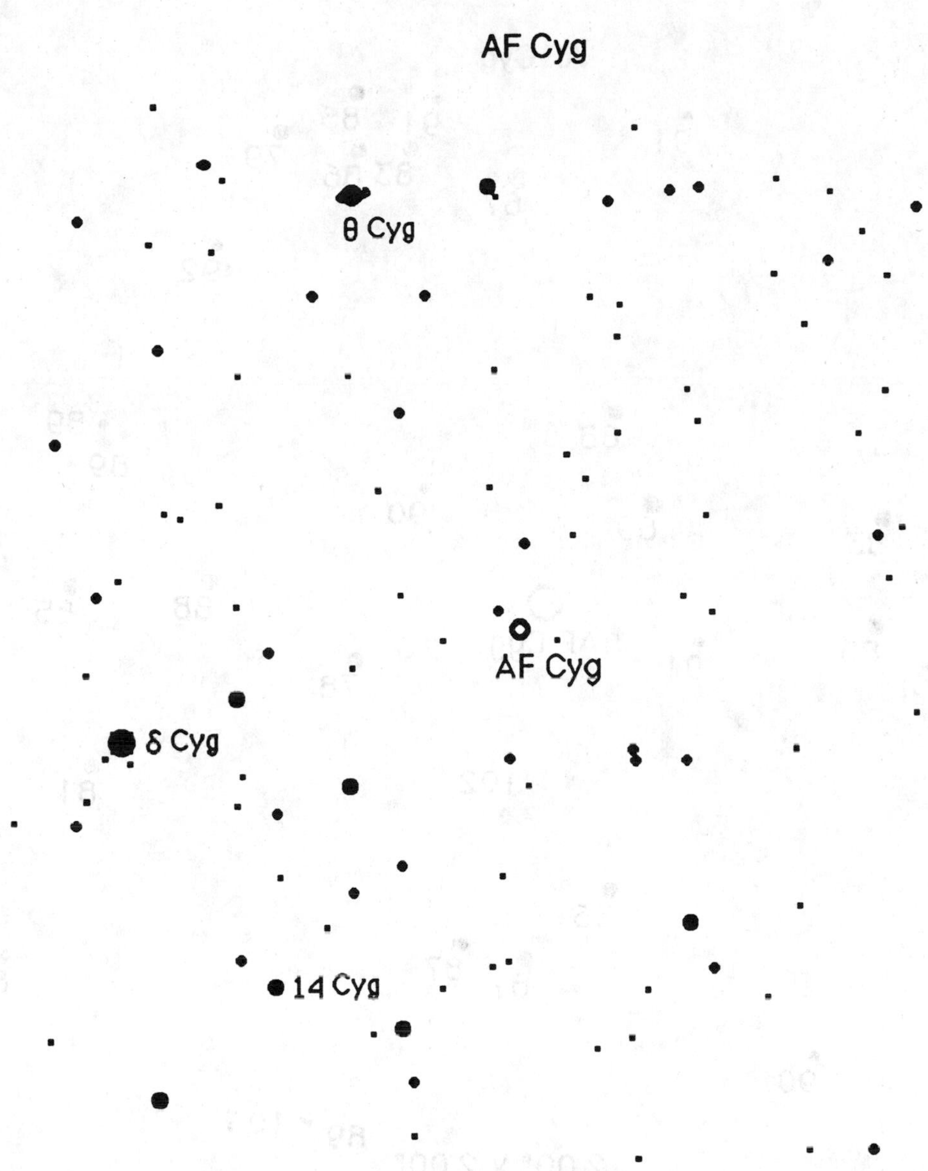

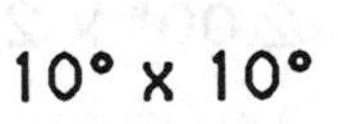

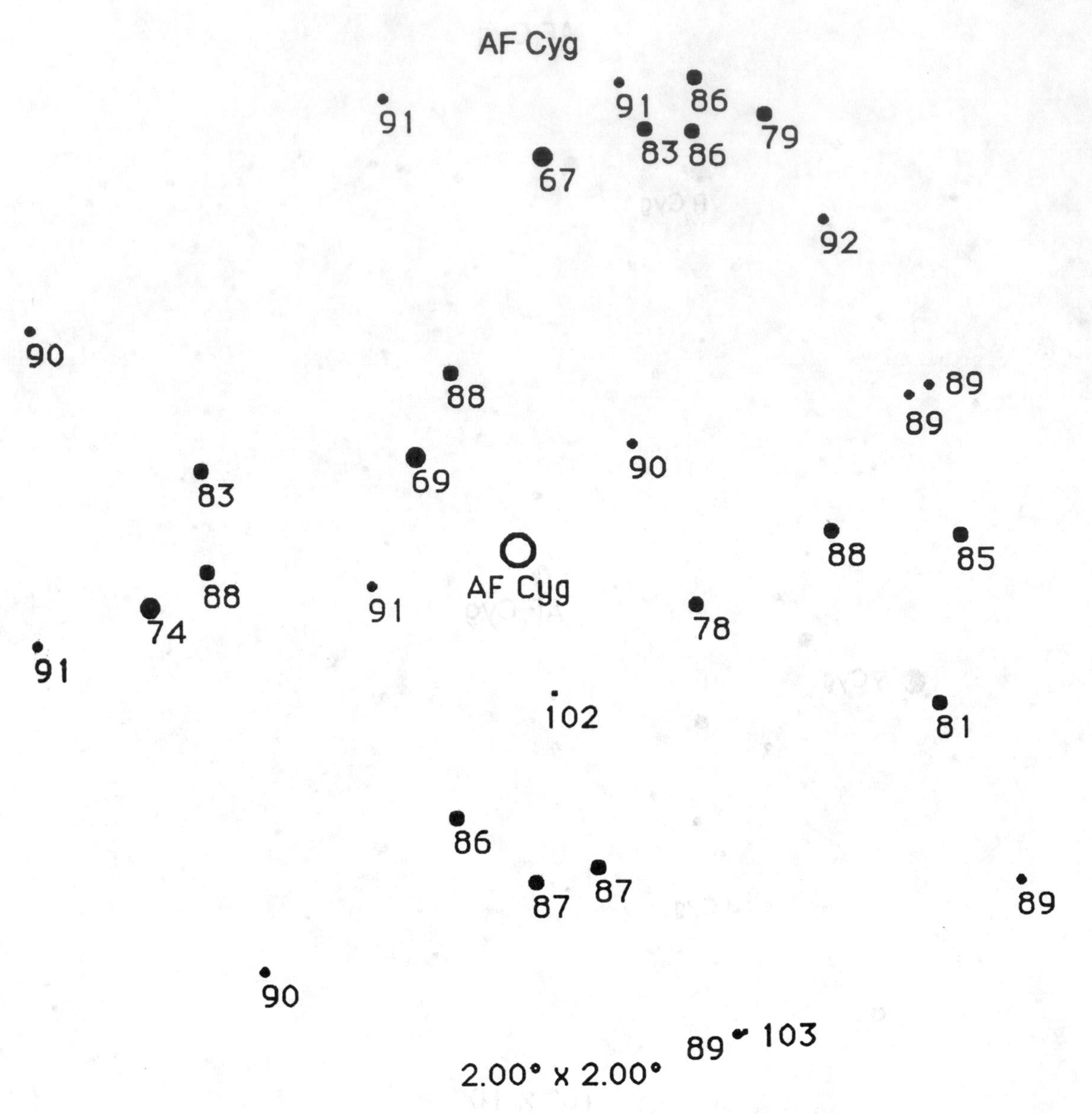
AF Cyg
91
91
86
83
86
79
67
92
90
88
89
89
83
69
90
88
85
AF Cyg
88
74
91
78
91
102
81
86
87
87
89
90
89
103
2.00° x 2.00°

Project A-6
Eclipsing Variable Stars

Eclipsing stars provide the special circumstance that lets you see stars orbit about each other in hours and days rather than years. By lucky coincidence, their orbits are oriented in space just perfectly for us to see each component star pass in front of its partner as they orbit about each other. An eclipsing binary is brightess when both stars are visible (outside eclipse) but fades to a minimum as each star passes in front of its partner and blocks its partner's light (during eclipse). The minima are often unequal — the deeper minima is called the primary minima, and the shallower minima is called the secondary minimum (Figure 1). Eclipsing binaries come in three types; EA, EB and EW.

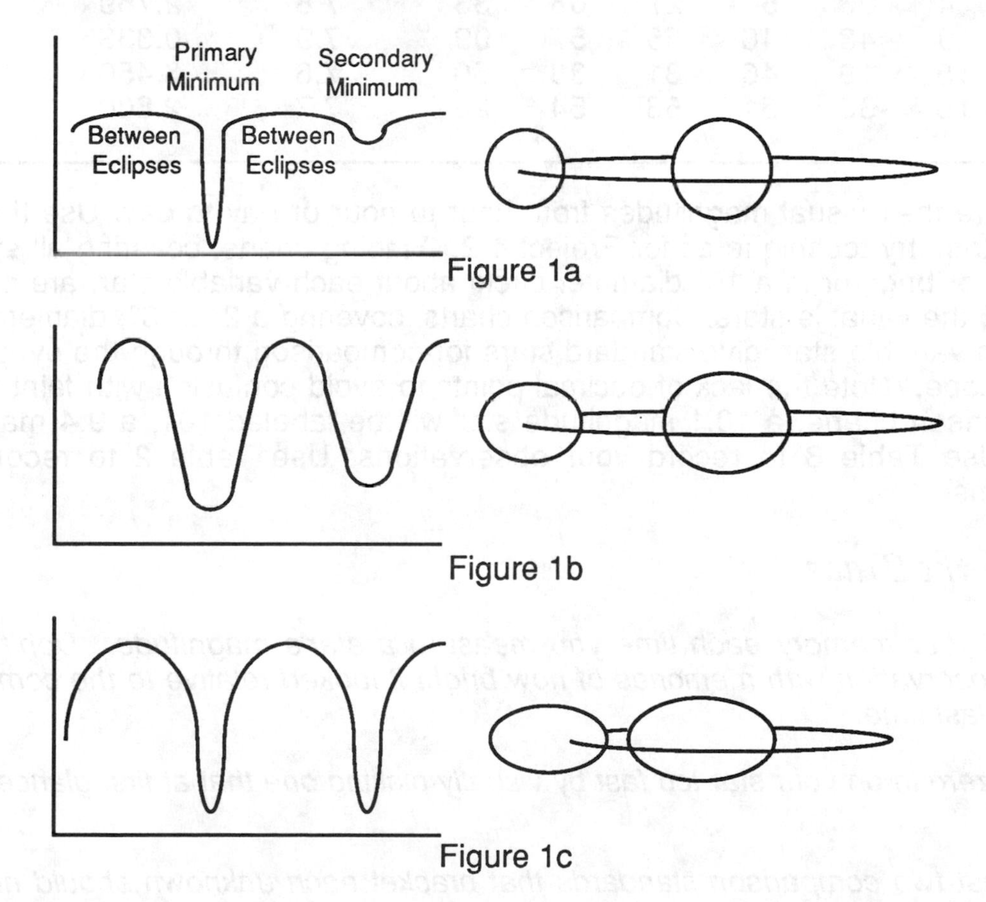

Figure 1a

Figure 1b

Figure 1c

The prototype star for type EA is Algol (β Per) which consists of two stars revolving about each other in well separated orbits. Eclipses show a deep primary minimum and a shallow secondary with the brightness outside the eclipses being fairly constant (Figure 1a).

The prototype for type EB is β Lyr. In EB binaries (Figure 1b) the two stars, usually giants or supergiants, are closer together than in EA binaries and their mutual gravitation distorts each others shape into ovoids. As a result their light curve is rounded and they the secondary minima can approaches the primary minima in depth.

The last type of eclipsing binary is EW, after W UMa. The two main-sequence stars composing a EW binary are very close to each other — sometimes in actual contact (Figure 1c). The primary and secondary minima are about the same depth.

Table 1 - Eclipsing Variable Stars

Name	RA			Dec			Peak Visual	Period	Variable
	h	m	s	°	'	"	Magnitude	(days)	Type
U Cep	1	02	18	81	52	32	6.8	2.493	EA
RW Tau	4	03	54	27	08	33	7.8	2.769	EA
W UMa	9	43	46	55	57	09	7.9	0.3336	EW
U CrB	15	19	46	31	38	50	7.6	3.450	EA
TW Dra	15	33	51	63	54	26	7.7	2.800	EA

Observe their visual magnitudes from hour to hour or day to day. Use the same visual photometry technique as for Project 4-2. Finding charts, covering all stars 8th magnitude or brighter in a 10° diameter circle about each variable star, are provided for locating the variable stars. Comparison charts, covering a 2° or 3° diameter circle about each variable star, give standard stars for comparison through the eyepiece of your telescope. Note the lack of decimal points to avoid confusion with faint stars in the comparison charts; a 10.1 magnitude star will be labeled 101, a 9.4 magnitude star 94. Use Table 3 to record your observations. Use Table 2 to record your observations.

Tricks of the Trade

1) Purge your memory each time you measure a star's magnitude. Don't pollute tonight's observation with memories of how bright it looked relative to the comparison standards last time.

2) Don't zero in on your star too fast by visually picking one that at first glance seems to match .

3) The last two comparison standards that bracket each unknown should not differ by more than 0.5 magnitudes.

4) Look for small patterns of stars (triangles, squares, pentagons, etc.) to find your way around .

5) Try using higher power eyepieces on the fainter stars.

Questions

1) How many magnitudes did each variable change by during the primary minimum?

2) How many magnitudes did each variable change by during the secondary minimum?

3) How long were the eclipses compared to the time between eclipses? Were there differences between the different types of eclipsing variables?

4) Is there any relationship between how long the variables took to vary (period) and the change in visual magnitude over the cycle?

5) Were there any color changes and, if so, did color correlate with the star's cycle of variation (i.e., between eclipses, primary minimum, secondary minimum)?

Table 3: Variable Observations

Obs Number	Date	Time	RA h	RA m	Dec °	Dec '	Alt °	Azi °	Visual Mag

U Cep

U Cep

10°x 10°

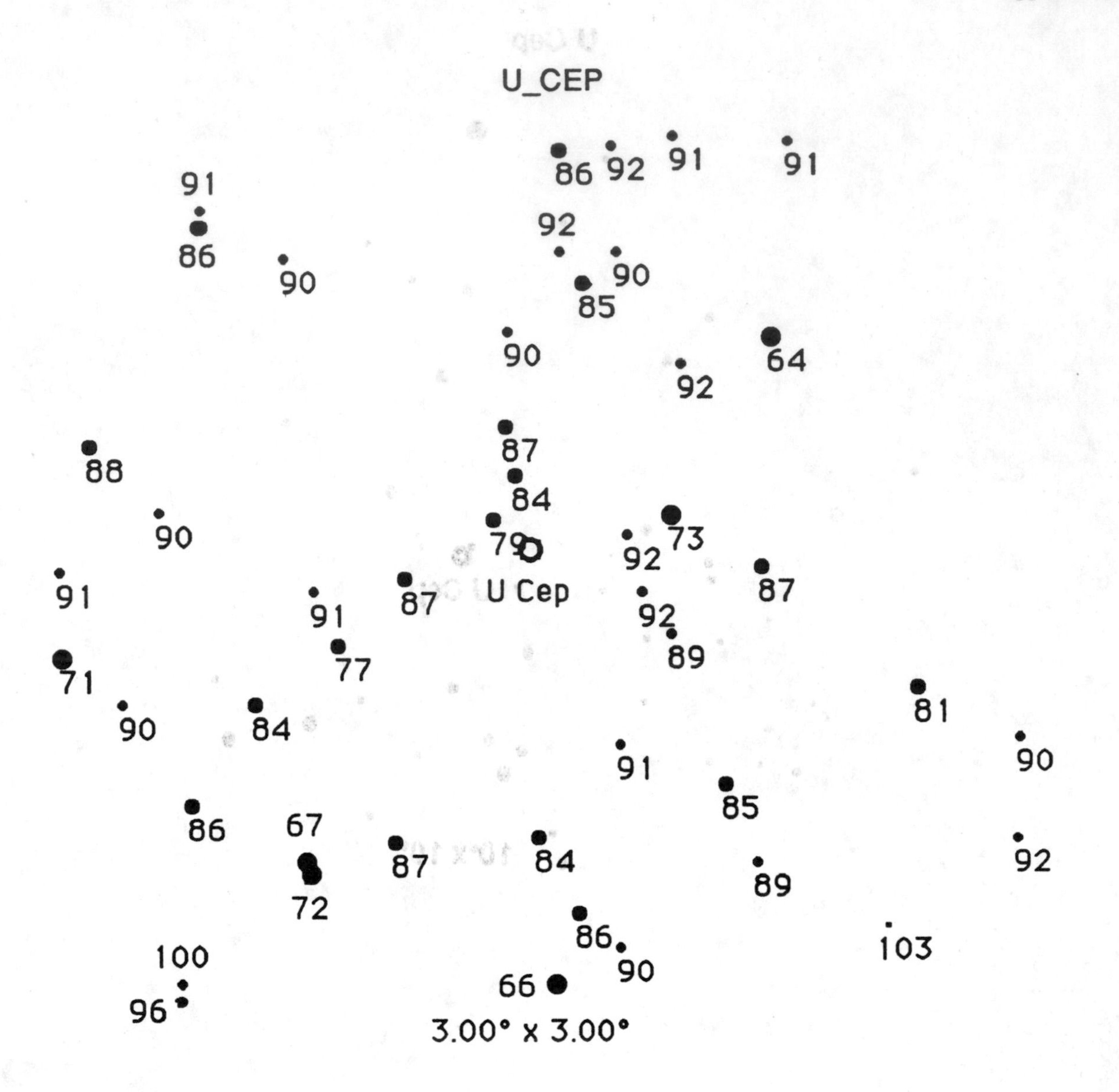
U_CEP
86
92
91
91
91
86
90
92
90
85
90
64
92
87
88
84
90
73
79
U Cep
92
87
91
91
87
92
89
77
71
81
90
84
90
91
85
86
67
87
84
92
89
72
86
103
100
90
66
96
3.00° x 3.00°

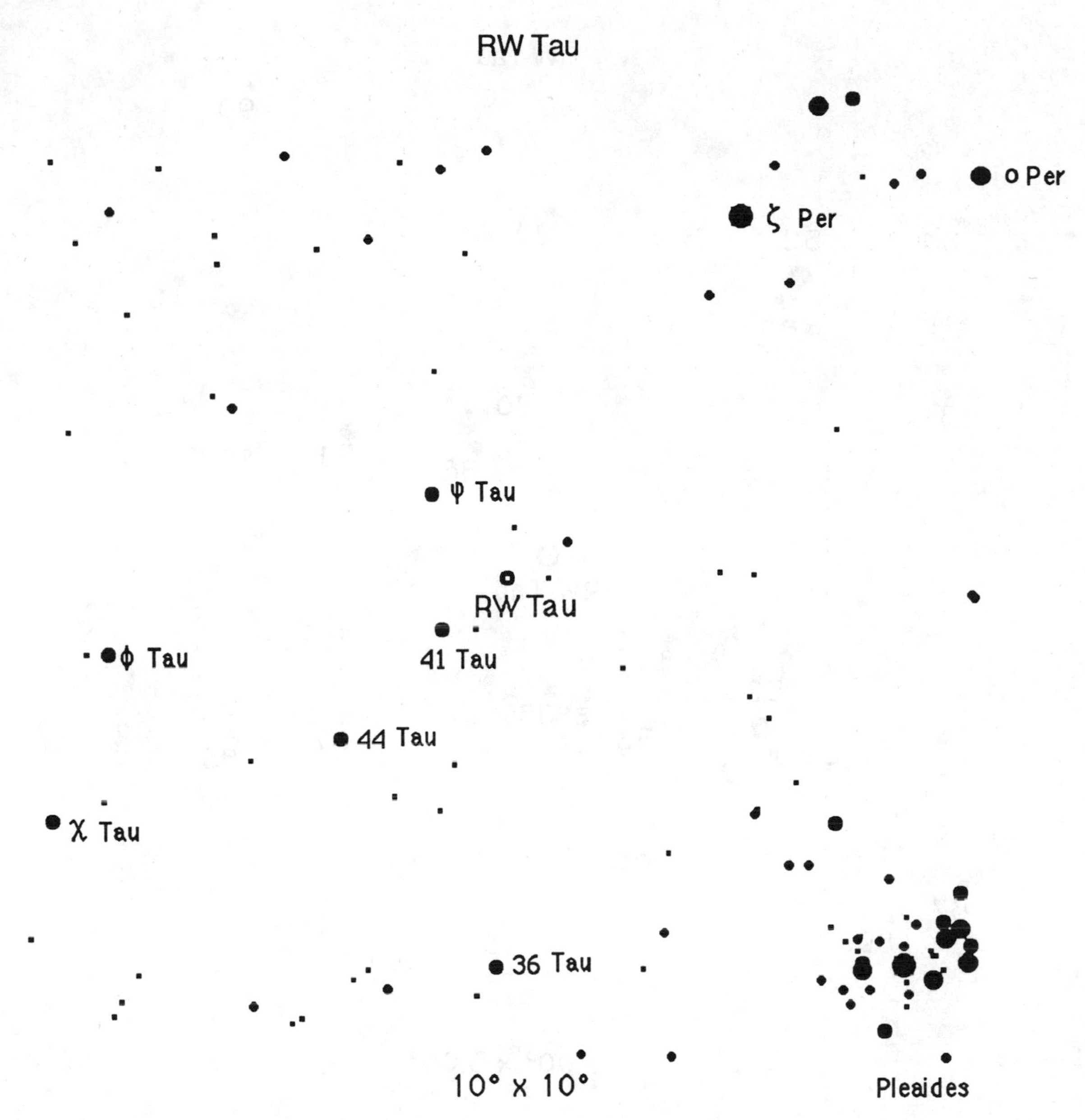
RW Tau
o Per
ζ Per
φ Tau
RW Tau
ϕ Tau
41 Tau
44 Tau
χ Tau
36 Tau
10° x 10°
Pleaides

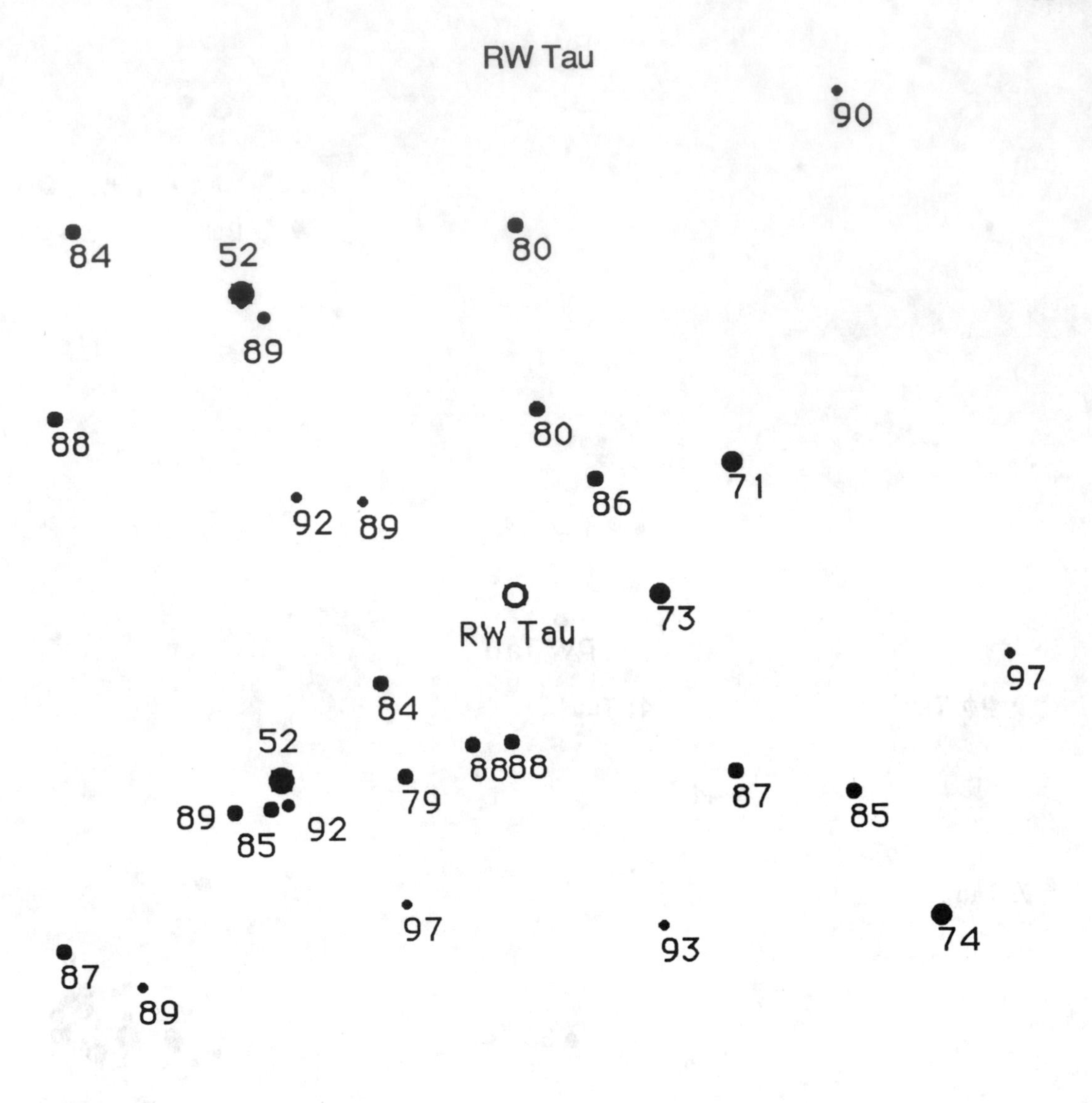
RW Tau
90
84
52
80
89
88
80
71
86
92
89
RW Tau
73
97
84
52
88
88
87
79
85
89
92
85
97
93
74
87
89
3.00° x 3.00°

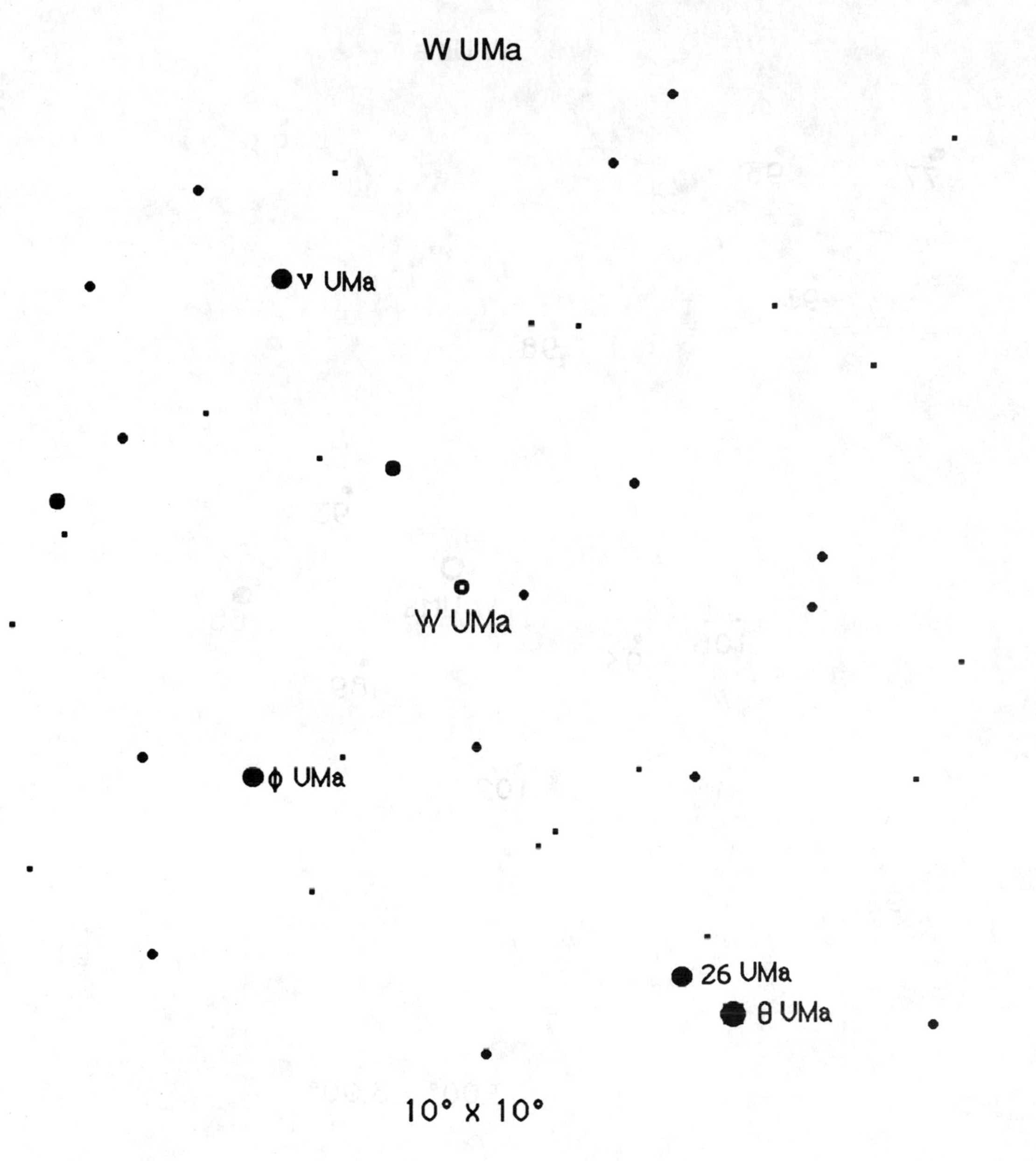
W UMa
ν UMa
W UMa
ϕ UMa
26 UMa
θ UMa
10° x 10°

W Uma

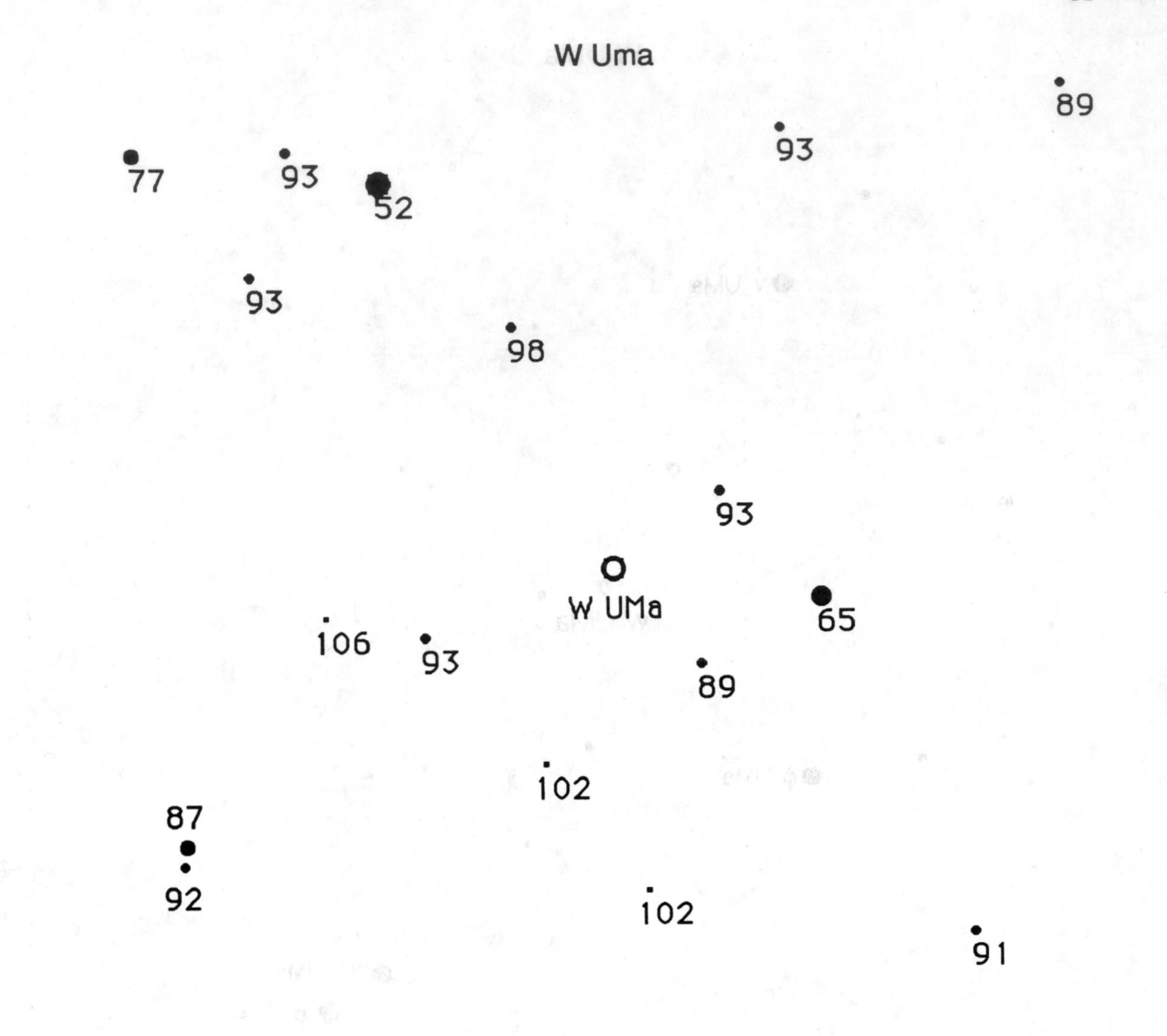

U CrB

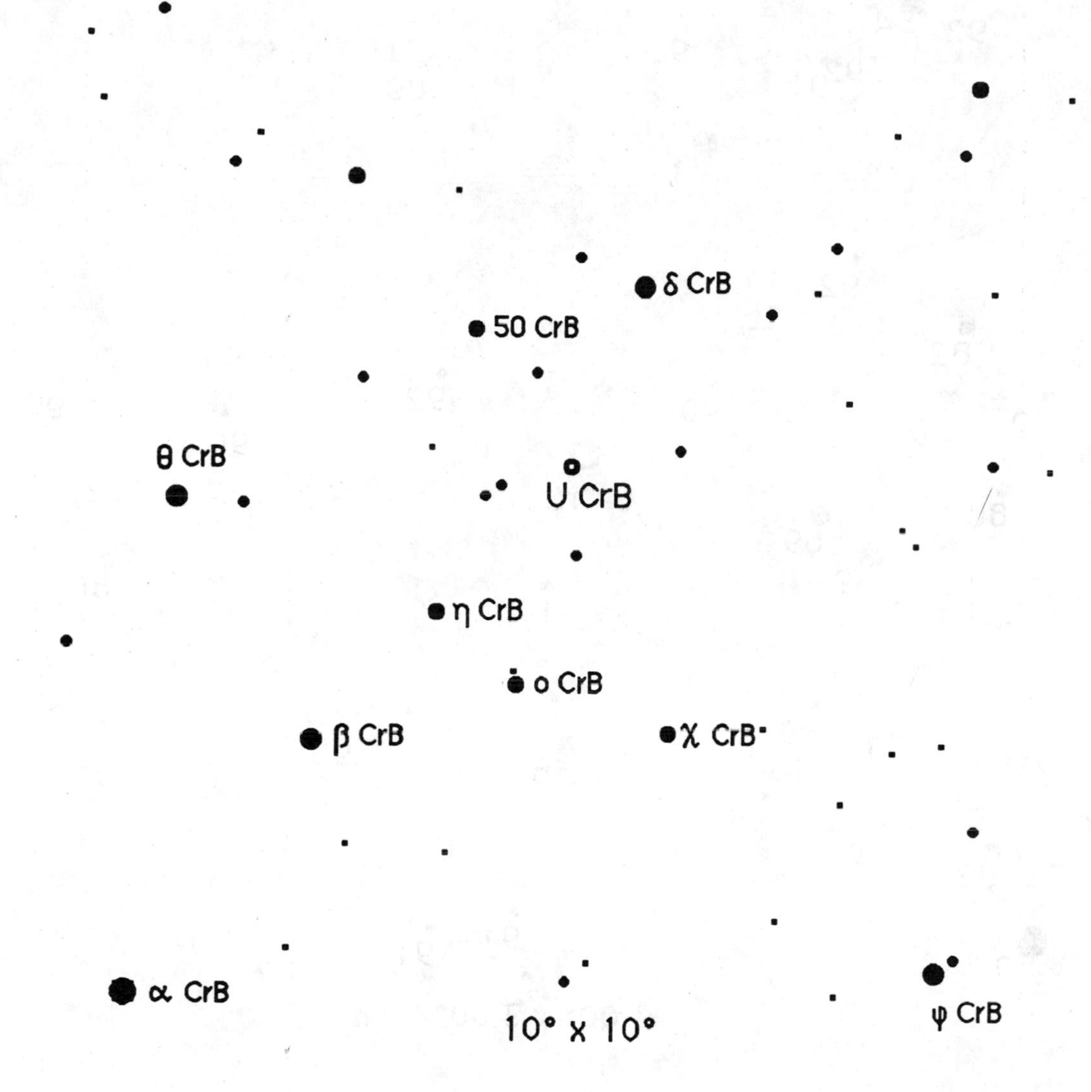
δ CrB
50 CrB
θ CrB
U CrB
η CrB
o CrB
β CrB
χ CrB
α CrB
φ CrB
10° x 10°

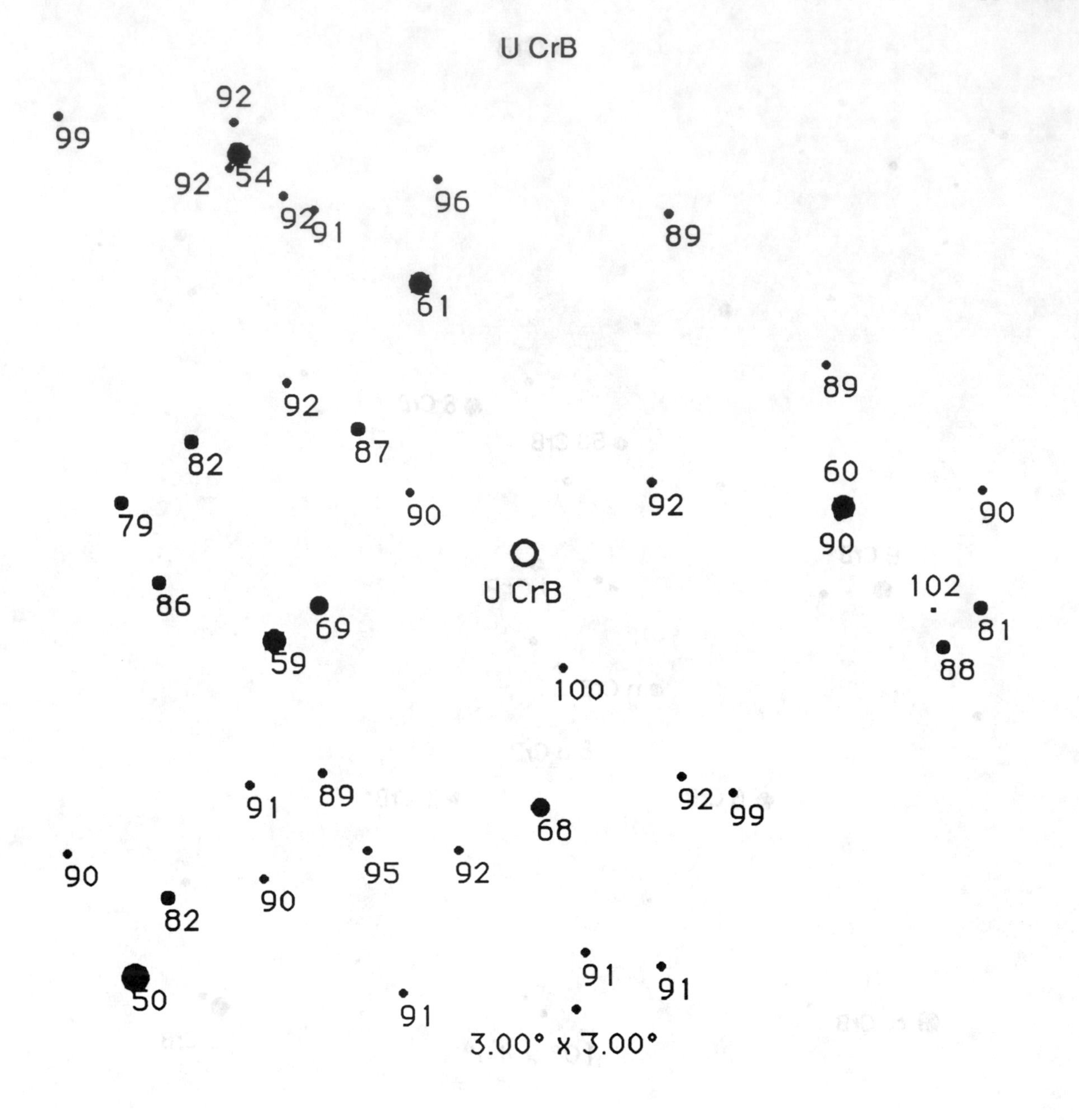
U CrB
92
99
92
54
92
91
96
89
61
89
92
87
82
60
90
90
92
79
90
U CrB
102
86
69
81
59
88
100
89
91
92
99
68
90
95
92
90
82
91
91
50
91
3.00° x 3.00°

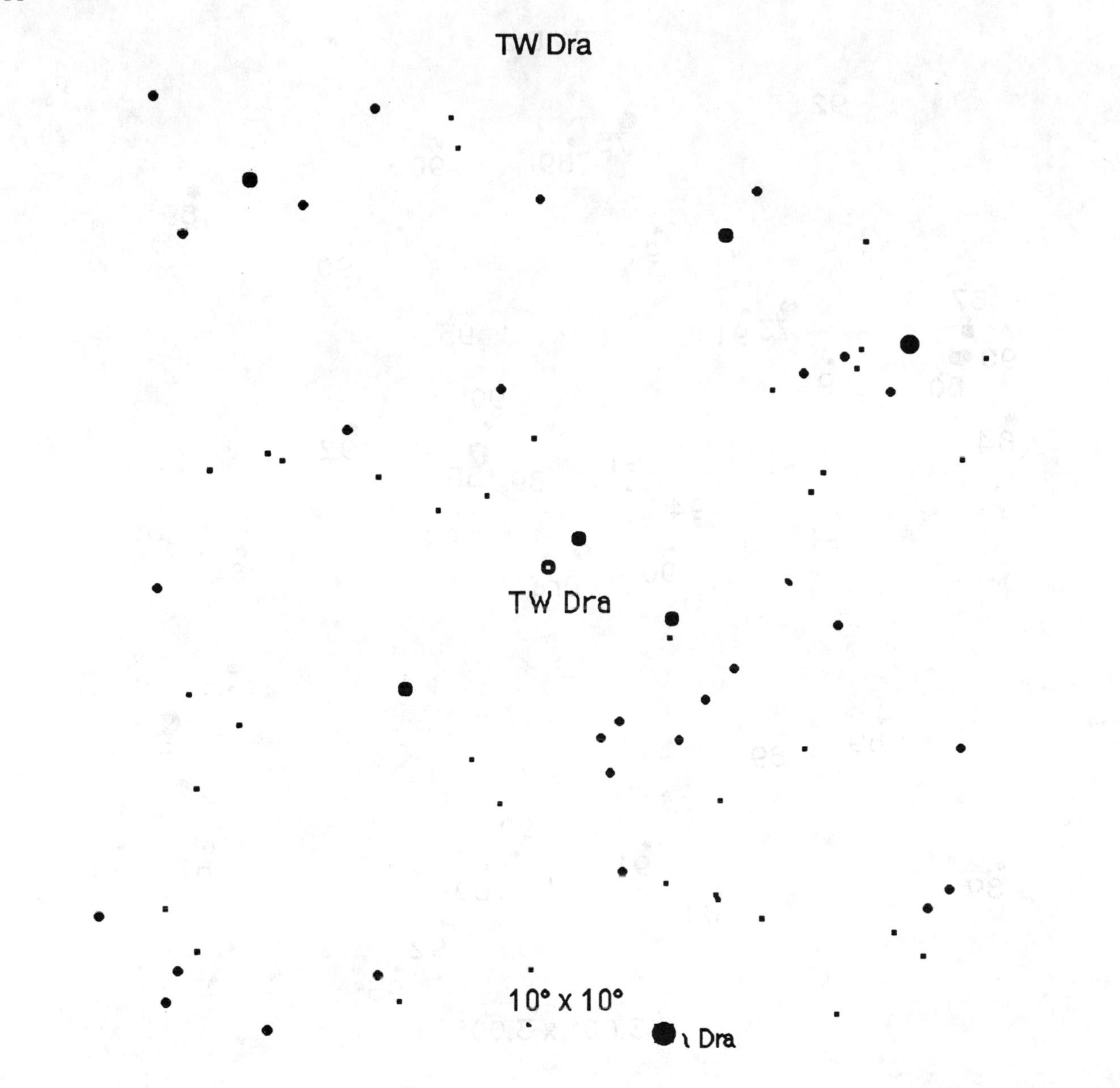
TW Dra
TW Dra
10° x 10°
ι Dra

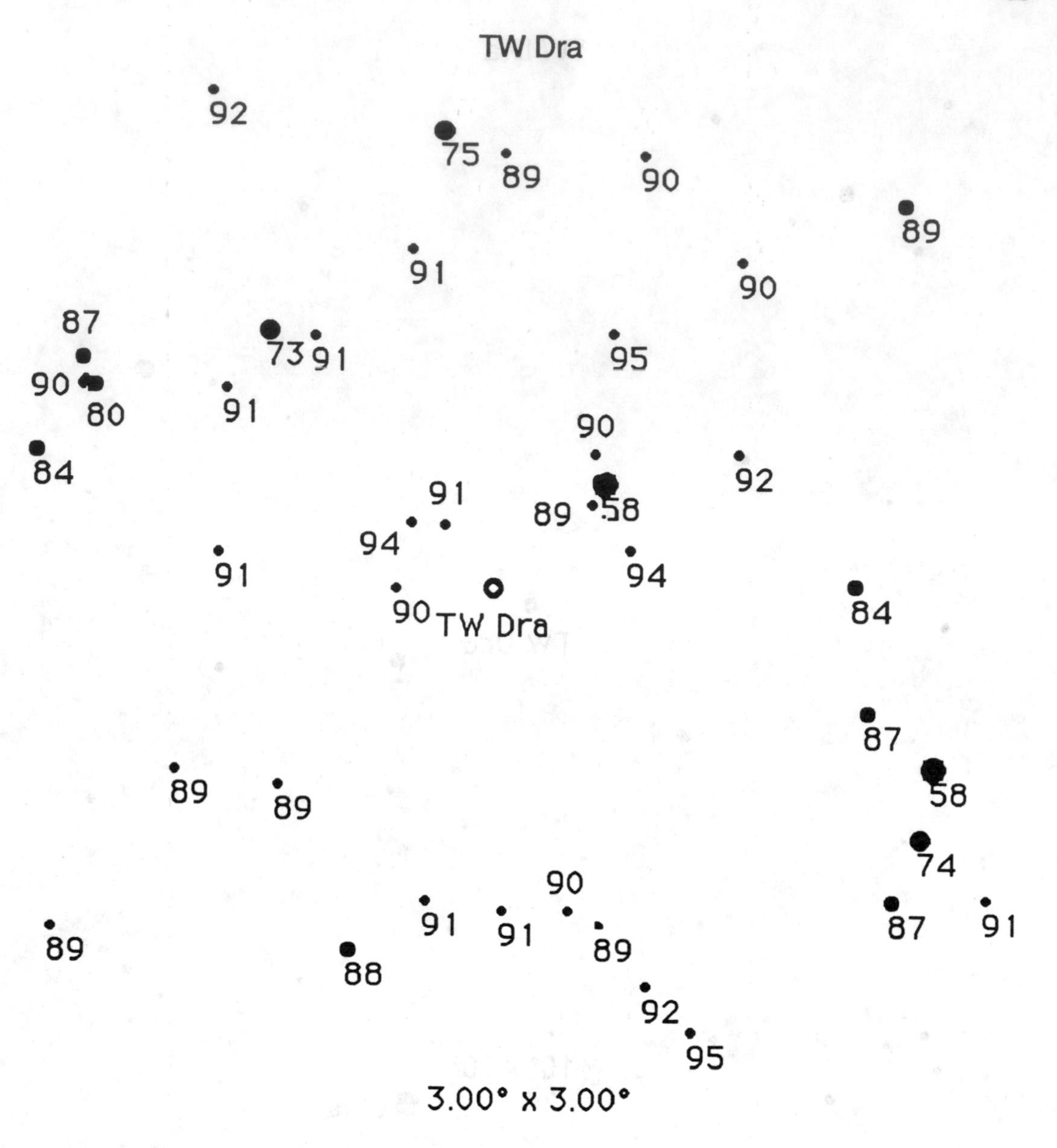
TW Dra
92
75
89
90
89
91
90
87
73
91
95
90
80
91
84
90
92
91
89
58
94
91
94
90
TW Dra
84
87
58
89
89
74
90
89
91
91
89
88
87
91
92
95
3.00° x 3.00°

Appendix B
Amateur Astronomy Clubs & Organizations

Alabama

Birmingham Astronomical Society
P.O. Box 36311
Birmingham, AL 35236
(205) 979-9343

Mobile Astronomical Society
P.O. Box 190042
Mobile, AL 36619
(205) 973-1325

Von Braun Astronomical Society, Inc.
Box 1142
Huntsville, AL 35807
(205) 881-0793

Arizona

Phoenix Astronomical Society
6945 E. Gary Road
Scottsdale, AZ 85254
(602) 996-3617

Tucson Amateur Astronomy Assn.
7222 E. Brooks Dr.
Tucson, AZ 85730
(602) 790-5053

Arkansas

Arkansas-Oklahoma Astronomical Society
P.O. Box 31
Fort Smith, AR 72902
(501) 452-4614

California

Central Coast Astronomical Society
P.O. Box 1415
San Luis Obispo, CA 93406
(805) 528-6682

Central Valley Astronomers, Inc.
5790 E. Tarpey Dr.
Fresno, CA 93727
(209) 291-7879

China Lake Astronomical Society
P.O. Box 1783
Ridgecrest, CA 93555
(619) 375-5681

Eastbay Astronomical Society
4917 Mountain Boulevard
Oakland, CA 94619
(415) 533-2394

Idyll-Gazers Astronomy Club
P.O. Box 1245
Idyllwild, CA 92349
(714) 659-3562

Los Angeles Astronomical Society
2800 E. Observatory Road
Los Angeles, CA 90027
(213) 926-4071

Oceanside Photo and Telescope Astronomical Society
929 Buena Rosa Ct.
Fallbrook, CA 92028
(619) 723-0684

Orange County Astronomers
2215 Martha Ave.
Orange, CA 92667
(714) 639-8446

Peninsula Astronomical Society
P.O. Box 4542
Mountainview, CA 94040
(415) 566-3116

Polaris Astronomical Society
22018 Ybarra Road
Woodland Hills, CA 91364
(818) 347-8922

Pomona Valley Amateur Astronomers
546 Prospesctors Road
Diamond Bar, CA 91765
(714) 860-5373

Riverside Astronomical Society
P.O. Box 7213
Riverside, CA 92503
(714) 689-6893

Sacramento Valley Astronomical Society
P.O. Box 575
Rocklin, CA 95677
(916) 624-3333

San Bernardino Valley Amateur Astronomers
1345 Garner Ave.
San Bernardino, CA 92411

San Diego Astronomy Association
P.O. Box 23215
San Diego, CA 92123
(619) 587-0172

San Francisco Amateur Astronomers
114 Museum Way
San Francisco, CA 94114
(415) 752-9420

San Jose Astronomical Association
3509 Calico Ave.
San Jose, CA 95124
(408) 371-1307

Sonoma County Astronomical Society
P.O. Box 183
Santa Rosa, CA 95404
(707) 528-1034

Stockton Astronomical Society
P.O. Box 243
Stockton, CA 95201
(209) 473-8234

Tulare Astronomical Association, Inc.
P.O. Box 515
Tulare, CA 93275
(209) 685-0585

Ventura County Astronomical Society, Inc.
P.O. Box 982
Simi Valley, CA 93063

Colorado

Denver Astronomical Society
P.O. Box 10814
Denver, CO 80210

Rocky Mountain Astrophysical Group
P.O. Box 25233
Colorado Springs, CO 80936
(719) 550-9804

Connecticut

Fairfield County Astronomical Society
Stamford Museum Observatory
39 Scofieldtown Road
Stamford, CT 06903
(203) 322-1648

Mattatuck Astronomical Society
Mattatuck Community College
Math-Science Division
750 Chase Parkway
Waterbury, CT 06708

Westport Astronomical Society
P.O. Box 5118
Westport, CT 06880

Delaware

Delaware Astronomical Society
P.O. Box 652
Wilmington, DE 19899
(215) 444-2966

Florida

Central Florida Astronomical Society
810 E. Rollins Street
Orlando, FL 32803
(305) 323-8890

Escambia Amateur Astronomers Association
6235 Omie Circle
Pensacola, FL 32504
(904) 484-1154

Local Group of Deep Sky Observers
2311 23rd Ave. W.
Bradenton, FL 34205
(813) 747-8334

St. Petersburg Astronomical Club, Inc.
594 59th Street S.
St. Petersburg, FL 33707
(813) 343-1594

Georgia

Atlanta Astronomy Club
5198 Avanti Ct.
Stone Mountain, GA 30088
(404) 498-1240

Oglethorpe Astonomical Association
Savannah Science Museum
4405 Paulsen Street
Savannah, GA 31405
(912) 355-6705

Hawaii

Hawaiian Astronomical Society
P.O. Box 17671
Honolulu, HI 96817

Maui Astronomy Club
325 Olokani Street
Makawao, Maui , HI 96768
(808) 572-1939

Mauna Kea Astronomical Society
R. R. #1, Box 525
Captain Cook, HI 96704
(808) 328-9201

Idaho

Boise Astronomical Society
10879 Ashburton Dr.
Boise, ID 83709
(208) 377-5220

Illinois

Astronomical Society at University of Illinois
349 Astronomy Bldg.
1011 W. Springfield
Urbana, IL 61801
(217) 351-7898

Chicago Astronomical Society
P.O. Box 48504
Chicago, IL 60648
(312) 966-6214

Naperville Astronomical Association
205 N. Mill Street
Naperville, IL 60540
(312) 355-5357

Northwest Suburban Astronomers
4960 Chambers Dr.
Barrington, IL 60010

Peoria Astronomical Society, Inc.
1125 W. Lake Ave.
Peoria, IL 61604
(309) 347-7285

Twin City Amateur Astronomers
P.O. Box 755
Normal, IL 61761
(309) 454-4164

Indiana

Evansville Astronomical Society, Inc.
P.O. Box 3474
Evansville, IN 47733
(812) 922-5681

Fort Wayne Astronomical Society
P.O. Box 6004
Fort Wayne, IN 46896
(219) 747-0774

Indiana Astronomical Society
2 Wilson Dr.
Mooresville, IN 46158
(317) 831-8387

Iowa

Des Moines Astronomical Society, Inc.
2307 49th Street
Des Moines, IA 50310
(515) 274-1873

Quad Cities Astronomical Society
P.O. Box 3706
Davenport, IA 52808
(319) 324-4661

Kansas

Kansas Astronomical Observers
220 S. Main
Wichita, KS 67202
(316) 264-3174

Kentucky

Blue Grass Amateur Astronomy Society
1490 N. Forbes Road
Lexington, KY 40511
(606) 252-6143

Louisiana

Pontchartrain Astronomy Society, Inc.
1441 Avenue A
Marrero, LA 70072
(504) 340-0256

Shreveport Astronomical Society, Inc.
1426 Alma Street
Shreveport, LA 71108
(318) 865-2433

Maryland

Baltimore Astronomical Society
601 Light Street
Baltimore, MD 21230
(301) 766-6605

Harford County Astronomical Society
P.O. Box 906
Bel Air, MD 21014
(301) 457-5597

Westminster Astronomical Society
3481 Salem Bottom Road
Westminster, MD 21157
(301) 848-6384

Massachusetts

Amateur Telescope Makers of Boston, Inc.
8 Pond Street
Dover, MA 02030
(508) 785-0352

South Shore Astronomical Soiciety
P.O. Box 429
Jacobs Lane
Norwell, MA 02061
(617) 588-0673

Michigan

Detroit Astronomical Society
14298 Lauder
Detroit, MI 48227
(313) 981-4096

Grand Rapids Amateur Astronomical Association
4 Alten N.E.
Grand Rapids, MI 49503
(616) 454-7645

Warren Astronomical Society
P.O. Box 474
East Detroit, MI 48021
(313) 355-5844

Minnesota

Minnesota Astronomical Society
30 E. 10th Street
St. Paul , MN 55101
(612) 451-7680

3M Astronomical Society
14601 55th Street S.
Afton, MN 55001
(612) 733-2690

Mississippi

Jackson Astronomical Association
6207 Winthrop Circle
Jackson, MS 39206
(601) 982-2317

Missouri

Astronomical Society of Kansas City
P.O. Box 400
Blue Springs, MO 64015
(816) 228-4238

St. Louis Astronomical Society, Inc.
4562 Clearbrook Dr.
St. Charles, MO 63303

Montana

Astronomical Institute of the Rockies
6351 Canyon Ferry Road
Helena, MT 59601
(406) 442-2208

Nebraska

Omaha Astronomical Society
5025 S. 163 Street
Omaha, NE 68135
(402) 896-4417

Prairie Astronomy Club
P.O. Box 80553
Lincoln, NE 68501
(402) 467-4222

Nevada

Astronomical Society of Nevada
825 Wilkinson Ave.
Reno, NV 89502
(702) 329-9946

Las Vegas Astronomical Society
Clark County Community College Planetarium
3200 E. Cheyenne Ave
Las Vegas, NV 89030
(702) 459-8401

New Hampshire

New Hampshire Astronomical Society
22 Center Street
Penacook, NH 03303
(603) 753-9225

New Jersey

Amateur Astronomers Association of Princeton, Inc.
P.O. Box 2017
Princeton, NJ 08540
(609) 396-3630

Amateur Astronomers, Inc.
W. M. Spezzy Observatory
1033 Springfield Ave.
Cranford, NJ 07016
(201) 549-0615

New Jersey Astronomical Association
Voorhees State Park
P.O. Box 214
High Bridge, NJ 08829
(215) 253-7294

Small Scope Observers Association
4 Kingfisher Place
Audubon Park, NJ 08106
(609) 547-9487

New Mexico

Albuquerque Astronomical Society
P.O. Box 54072
Albuquerque, NM 87153
(505) 299-0891

Astronomical Society of Las Cruces
P.O. Box 921
Las Cruces, NM 88004
(505) 526-2968

New York

Amateur Astronomers Association of New York, Inc.
1010 Park Ave.
New York, NY 10028
(212) 535-2922

Astronomical Society of Long Island, Inc.
1011 Howells Road
Bay Shore, Long Island, NY 11706
(516) 586-1760

Broome County Astronomical Society
Roberson-Kopernik Observatory
Underwood Road
Vestal, NY 13850
(607) 748-3685

Buffalo Astronomical Association, Inc.
Buffalo Museum of Science
Humbolt Parkway
Buffalo, NY 14211

Rockland Astronomy Club
110 Pascack Road
Pearl River, NY 10965
(914) 735-4163

Syracuse Astronomical Society, Inc.
1115 E. Colvin Street
Syracuse, NY 13210
(315) 458-1454

Westchester Astronomy Club
511 Warburton Ave.
Yonkers, NY 10701
(914) 963-4550

North Carolina

Forsyth Astronomical Society
504 Gayron Drive
Winston-Salem, NC 27105
(919) 744-7141

Raleigh Astronomy Club
P.O. Box 10643
Raleigh, NC 27605
(919) 832-NOVA

North Dakota

Dakota Astronomical Society
P.O. Box 2539
Bismarck, ND 58502
(701) 256-3620

Ohio

Astronomy Club of Akron, Inc.
5070 Manchester Road
Akron, OH 44319
(216) 644-0230

Cincinnati Astronomical Society
5274 Zion Road
Cleves, OH 45002
(513) 661-3252

Columbus Astronomical Society, Inc.
P.O. Box 16209
Coloumbus, OH 43216
(614) 262-9713

Miami Valley Astronomical Society
Dayton Museum of Natural History
2629 Ridge Ave.
Dayton, OH 45414
(513) 275-7431

Ohio Turnpike Astronomers Association
1494 Lakeland Ave.
Lakewood, OH 44107
(216) 521-5115

Oklahoma

Astronomy Club of Tulsa
P.O. Box 470611
Tulsa, OK 74147
(918) 742-7577

Oklahoma City Astronomy Club
2100 N.E. 52nd Street
Oklahoma City, OK 73111
(405) 424-5545

Oregon

Eugene Astronomical Society
Lane E. S. D. Planetarium
P.O. Box 2680
Eugene, OR 97402
(503) 741-0501

Portland Astronomical Society
2626 S.W. Luradel Street
Portland, OR 97219
(503) 245-6251

Pennsylvania

Amateur Astronomers Association of Pittsburgh, Inc.
Wagman Observatory
P.O. Box 314
Glenshaw, PA 15116
(412) 224-2510

Astronomical Society of Harrisburg
1915 Enfield Street
Camp Hill, PA 17011
(717) 975-9799

Delaware Valley Amateur Astronomers
6233 Castor Ave.
Philadelphia, PA 19149
(215) 831-0485

Lackawanna Astronomical Society
1112 Fairview Road
Clarks Summit, PA 18411
(717) 586-0789

Lehigh Valley Amateur Astronomical Society, Inc.
620 E. Rock Road
Allentown, PA 18103
(215) 398-7295

Rhode Island

Celestial Observers
Astronomy Department
P.O. Box 1843
Brown University
Providence, RI 02912
(401) 863-2046

Skyscrapers, Inc.
47 Peeptoad Road
North Scituate, RI 02857
(401) 942-7893

South Carolina

Carolina Skygazers Astronomy Club
Museum of York County
4621 Mount Gallant Road
Rock Hill, SC 29730
(803) 329-2121

South Dakota

Black Hills Astronomy Society
3719 Locust
Rapid City, SD 57701

Tennessee

Barnard Astronomical Society
P.O. Box 90042
Chattanooga, TN 37412
(615) 629-6094

Smoky Mountain Astronomical Society
P.O. Box 6204
Knoxville, TN 37914
(615) 637-1121

Texas

Austin Astronomical Society
P.O. Box 12831
Austin, TX 78711

Fort Worth Astronomical Society, Inc.
P.O. Box 161715
Fort Worth, TX 76161
(817) 860-6858

JSC Astronomical Society
3702 Townes Forest
Friendswood, TX 77546
(713) 482-3909

San Antonio Astronomical Association
6427 Thoreau's Way
San Antonio, TX 78239
(512) 654-9784

Texas Astronomical Society
P.O. Box 25162
Dallas, Tx 75225
(214) 368-6982

Utah

Salt Lake Astronomical Society
15 S. State Street
Salt Lake City, UT 84111
(801) 538-2104

Virginia

Northern Virginia Astronomy Club
6028 Ticonderoga Ct.
Burke, VA 22015
(703) 866-4985

Richmond Astronomical Society
709 Timken Dr.
Richmond, VA 23239
(804) 741-3689

Washington

Seattle Astronomical Society
852 N.W. 67th Street
Seattle, WA 98115
(205) 523-2787

Spokane Astronomical Society, Inc.
4140 Cook Street
Spokane, WA 99223
(509) 448-9694

Tacoma Astronomical Society
7101 Topaz Dr. S.W.
Tacoma, WA 98498
(206) 588-9504

Wisconsin

Milwaukee Astronomical Society
W248 S7040 Sugar Maple Dr.
Waukesha, WI 53186
(414) 662-2987

Wyoming

Cheyenne Astronomical Society
3409 Frontier Street
Cheyenne, WY

Appendix C
Trained Eye Star Atlas

Multiple Stars

Bayer-Flamsteed Name	RA h	RA m	Dec °	Dec '	Visual Magnitudes				Seperations "		
35 Psc	0	15	8	49	5.9	7.6			11.0		
β1, β2 Tuc	0	32	-62	58	4.4	13.5	4.8	6.0	2.4	27.1	0.5
π And	0	37	33	43	4.4	8.6			35.9		
55 Psc	0	40	21	26	5.4	8.7			6.5		
η Cas	0	49	57	49	3.5	7.3			12.0		
ψ1 Psc	1	6	21	28	5.6	5.8			30		
ζ Psc	1	14	7	34	5.6	5.8			23.6		
37 Cet	1	14	-7	55	5.2	8.7			49.7		
ψ Cas	1	26	68	8	4.7		9.6	9.7	25.0	2.9	
ε Scl	1	40	025	3	5.4	8.6			4.7		
γ Ari	1	54	19	18	4.8	4.8			7.8		
λ Ari	1	58	23	35	4.8	7.1			37.5		
γ And	2	04	42	20	2.2	5.1			9.8		
6 Tri	2	12	30	18	5.4	7.0			3.9		
66 Cet	2	13	-2	24	5.7	7.5			16.5		
α UMi	2	20	89	16	2.0	8.9			18.4		
ι Cas	2	29	67	24	4.7	7.0		8.3	2.5	7.3	
ω For	2	34	-28	14	5.0	7.7			10.8		
30 Ari	2	37	24	38	6.6	7.3			38.0		
η Per	2	51	55	53	3.9	8.6			8.2		
α Cet	3	02	4	05	2.5	5.6			960.0		
a For	3	12	-28	59	4.0	6.6			5.1		
ζ Per	3	54	31	53	2.9	9.4			12.9		
32 Eri	3	54	-2	58	5.0	6.3			6.9		

Bayer-Flamsteed Name	RA h	RA m	Dec °	Dec '	Visual Magnitudes				Seperations "		
ε Per	3	58	40	00	3.0	8.1			9.0		
39 Eri	4	14	-10	15	5.0	8.0			6.4		
o^2 Eri	4	15	-7	39	4.4		9.5	11.2	83.4	9.2	
χ Tau	4	23	25	38	5.5	7.6			19.4		
α Tau	4	36	16	31	0.9	13.4			121.7		
ι Pic	4	51	-53	28	5.6	6.4			12.3		
ρ Ori	5	13	2	52	4.5	8.3			7.0		
β Ori	5	14	-8	11	0.2	7.0			9.2		
δ Ori	5	32	-0	22	2.5	6.9			52.8		
λ Ori	5	35	-9	52	3.7	5.6			4.4		
θ Ori	5	35	-5	54	5.1	6.7	6.7	8.0	13	17	13
σ Ori	5	39	-2	36	4.0	7.0	7.0	6.0	13	42	11
g Lep	5	45	-22	27	3.7	6.3			96.3		
θ Aur	6	00	37	13	2.6	7.1			3.4		
ε Mon	6	24	4	34	4.5	6.5			13.2		
γ Lep	5	44	-22	26	3.8	6.4			96.0		
ε Mon	6	24	4	36	4.5	6.5			13.4		
β Mon	6	29	-7	02	4.7		5.2	6.1	7.3	2.8	
20 Gem	6	32	17	47	6.3	6.9			20.0		
ν^1 CMa	6	36	-18	40	5.8	8.5			17.5		
α CMa	6	45	-16	43	-1.5	8.3			4.6		
12 Lyn	6	46	59	27	5.4	6.0		7.3	1.7	8.7	
38 Gem	6	55	13	11	4.7	7.7			5.2		
ε CMa	6	59	-28	58	1.6	8.1			7.4		
γ Vol	7	9	-70	30	4.0	5.9			13.6		
λ Gem	7	18	16	32	3.6	10.7			9.6		
δ Gem	7	20	22	00	3.5	8.1			6.1		

Bayer-Flamsteed Name	RA h	RA m	Dec °	Dec '	Visual Magnitudes			Seperations "	
σ Pup	7	29	-43	18	3.3	9.4		22.3	
α Gem	7	35	31	54	2.0	2.9		2.2	
κ Gem	7	44	24	24	3.6	8.1		7.1	
γ Vel	8	9	-47	20	1.9	4.2	8.0	41.0	62.0
ζ Cnc	8	12	17	39	5.6	6.0	6.2	0.8	6.0
h^2 Pup	8	14	-40	21	4.4	9.5		51.1	
ι Cnc	8	47	28	46	4.2.	6.6		30.4	
σ^2 UMa	8	21	72	24	6.1	9.1		43.1	
ζ^1 Ant	9	31	-31	53	6.2	7.1		8.0	
23 Uma	9	31	63	04	3.7	8.9		22.7	
α Leo	10	08	11	58	1.3	7.6		176.5	
γ Leo	10	20	19	52	2.2	3.5		4.3	
I Vel	10	21	-56	03	4.7	8.4		7.2	
δ Ant	10	30	-30	36	5.6	9.6		11.0	
54 Leo	10	56	24	46	4.5	6.5		6.6	
γ Crt	11	25	-17	41	4.1	9.6		5.2	
2 CnV	12	16	40	39	5.8	8.9		11.5	
δ Crv	12	30	-16	30	3.0	8.4		24.2	
24 Com	12	35	18	22	5.2	6.7		20.3	
α CnV	12	56	38	18	2.9	5.4		19.7	
J Cen	13	23	-60	59	4.7	6.5		60.0	
ζ Uma	13	24	54	56	2.3	3.9		14.4	
Q Cen	13	42	-54	34	5.3	6.7		5.3	
3 Cen	13	52	-33	0	4.5	6.0		7.9	
4 Cen	13	53	-31	57	4.7	8.4		14.9	
α Cen	14	40	-60	50	0.0	1.3		14.1	
α Cir	14	43	-64	59	3.2	8.6		15.7	

Bayer-Flamsteed Name	RA h	RA m	Dec °	Dec '	Visual Magnitudes				Seperations "		
ε Boo	14	45	27	04	2.7	5.1			2.9		
α Lib	14	51	-16	02	2.9	5.3			231.0		
ξ Boo	14	51	19	05	4.8	6.9			7.2		
κ Lup	15	12	-48	44	3.9	5.8			26.8		
5 Ser	15	19	1	46	5.1	10.1			11.2		
μ Boo	15	24	37	23	4.3		7.0	7.6	108.3	2.3	
ζ CrB	15	39	36	38	5.1	6.0			6.3		
η Lup	16	0	-38	24	3.6	7.8			15.0		
β Sco	16	5	-19	48	2.7	10.3		4.9	0.5	13.7	
κ Her	16	08	17	03	5.3	6.5			28		
ν Sco	16	12	-19	28	4.2	6.8	6.4	7.8	0.8	41.1	2.3
σ CrB	16	15	33	52	5.6	6.6			7.1		
η Dra	16	24	61	31	2.9	8.8			6.1		
α Sco	16	30	-26	25	1.2	6.5			2.9		
α Her	17	15	14	23	2.5	5.0			4.6		
39 Oph	17	18	-24	17	5.4	6.9			10.8		
ν Dra	17	32	55	11	4.9	4.9			61.9		
ψ^1 Dra	17	42	72	09	4.9	6.1			30.2		
μ Her	17	47	27	43	3.4	10.1			33.8		
95 Her	18	02	21	35	5.2	5.3			6.3		
70 Oph	18	05	2	32	4.2	5.9			2.2		
39 Dra	18	24	58	47	5.1	7.8			3.7		
κ CrA	18	33	-38	44	5.9	6.6			21.4		
ε Lyr	18	44	39	40	5.1	6.0	5.1	5.4	2.7	207.8	2.3
ζ Lyr	18	45	37	36	4.3	5.9			43.7		
β Lyr	18	50	33	22	3.4	8.6			45.7		
15 Aql	19	5	-4	1	5.5	7.2			38.0		

Bayer-Flamsteed Name	RA		Dec		Visual Magnitudes				Seperations		
	h	m	°	'					″		
β Cyg	19	31	27	57	3.1	5.1			34.8		
ψ Cyg	19	55	52	25	4.9	7.4			3.1		
θ Sge	20	10	20	55	6.5	9.0			11.9		
β Cap	20	21	-14	46	3.1	6.2			205.0		
49 Cyg	20	41	32	18	5.7	7.8			2.7		
52 Cyg	20	46	30	43	4.2	9.4			6.0		
γ Del	20	47	16	07	4.3	5.1			9.8		
61 Cyg	21	07	38	45	5.2	6.0			30.3		
β Cep	21	29	70	33	3.2	7.8			13.6		
ξ Cep	22	04	64	37	4.6	6.6			7.6		
41 Aqr	22	14	-21	04	5.6	7.1			5.0		
δ Cep	22	29	58	25	4	6.3			40.7		
8 Lac	22	36	39	38	5.7	6.5	9.3	10.5	22	48	81
ψ¹ Aqr	23	16	-9	5	4.5	10.3			49.6		
94 Aqr	23	19	-13	28	5.3	7.5			13.3		
107 Aqr	23	46	-18	41	5.7	6.7			6.6		
σ Cas	23	59	55	45	5.1	7.2			3.1		

Non-Stellar Objects

GC = Globular Cluster
P = Planetary Nebula
HII = Excited Hydrogen Nebula
G = Galaxy

OC = Open Cluster
SNR = Supernova Remmnant
R = Reflection Nebula

NGC Number	Messier Number	RA h	RA m	Dec °	Dec '	Visual Magnitude	Type
55		0	15	-39	14	8.0	G
104		0	24	-72	05	4	G C
205	110	0	40	41	41	8	G
221	32	0	43	40	52	9.5	G
224	31	0	43	41	16	5.0	G
246		0	47	-12	09	8.5	P
253		0	47	-25	18	7.5	G
288		0	53	-26	36	9	G C
362		1	03	-70	51	6	G C
457		1	19	58	20	8	O C
581	103	1	33	60	42	7.4	G C
598	33	1	34	30	39	6.5	G
628	74	1	37	15	47	11.2	G
650-1	76	1	42	51	34	9.0	P
752		1	58	37	40	6.5	O C
869		2	19	57	08	4.4	O C
1039	34	2	42	42	46	6.0	O C
1068	77	2	43	-0	01	10	G
Pleiades	45	3	47	24	07	1.4	O C
1535		4	14	-12	44	9.0	P
1851		5	14	-40	03	7.3	G C
1904	79	5	24	-24	31	8.4	G C
1912	38	5	29	35	50	7.0	O C
1952	1	5	35	22	01	8.4	SNR
1960	36	5	36	64	08	6.5	O C
1976	42	5	36	-4	50	4.0	H II
1982	43	5	36	-5	16	9.0	H II

NGC Number	Messier Number	RA h	RA m	Dec °	Dec '	Visual Magnitude	Type
2068	78	5	47	0	04	8.3	R
2070		05	39	-69	06	5.0	H II
2099	37	5	53	32	33	6.2	OC
2168	35	6	09	24	19	5.5	OC
2237-9		6	32	4	37	5.5	H II
2264		6	41	9	53	4.7	OC
2287	41	6	47	-20	46	5.0	OC
2323	50	7	3	-8	21	7.2	OC
2392		7	29	20	55	8.0	P
2403		7	37	65	36	8.9	G
2422	47	7	37	-14	28	5.0	OC
2437	46	7	42	-14	49	6.6	OC
2447	93	7	45	-23	53	6.0	OC
2477		7	52	-38	33	5.5	OC
2516		7	58	-60	52	3.0	OC
2548	48	8	14	-5	46	5.5	OC
2632	44	8	40	19	41	4.0	OC
2682	67	8	51	11	48	7.4	OC
2808		9	12	-64	52	5.7	GC
3031	81	9	56	69	03	8.0	G
3034	82	9	56	69	41	9.2	G
3114		10	03	-60	07	4.4	OC
3132		10	07	-40	25	8.2	P
3201		10	17	-46	24	6.75	GC
3242		10	25	-18	39	8.9	P
IC 2602		10	43	-64	24	1.6	OC
3351	95	10	44	11	42	11.5	G
3368	96	10	47	11	49	10.4	G
3372		10	45	-59	50	4	H II
3379	105	10	48	12	35	10.8	G
3532		11	60	-58	40	7.0	OC
3556	108	11	12	55	40	11.0	G
3587	97	11	15	55	01	12.0	P

NGC Number	Messier Number	RA h	RA m	Dec °	Dec '	Visual Magnitude	Type
3623	65	11	19	13	06	10.5	G
3627	66	11	20	13	00	9.9	G
3918		11	50	-57	11	8.0	P
3992	109	11	58	53	22	11.2	G
4192	98	12	14	14	54	11.4	G
4254	99	12	19	14	25	10.5	G
4258	106	12	19	47	18	10.2	G
4303	61	12	22	4	28	10.4	G
4321	100	12	23	15	49	10.8	G
4372		12	26	-72	41	7.8	GC
4374	84	12	25	12	53	10.9	G
4382	85	12	25	18	11	10.5	G
4406	86	12	26	12	57	10.9	G
4472	49	12	30	8	00	10.1	G
4486	87	12	31	12	23	10.7	G
4501	88	12	32	14	25	10.9	G
4552	89	12	36	12	33	11.3	G
4565		12	36	25	59	10.5	G
4569	90	12	37	13	10	11.2	G
4579	58	12	38	11	49	11	G
4590	68	12	40	-26	45	9.1	GC
4594	104	12	40	-11	37	8.2	G
4621	59	12	42	11	39	11.4	G
4649	60	12	44	11	33	10.6	G
4736	94	12	51	41	06	8.9	G
4755		12	54	-60	20	5.2	OC
4826	64	12	57	21	40	8.6	G
4833		13	00	-70	53	8.0	GC
5024	53	13	13	18	10	8.7	GC
5055	63	13	16	42	02	10.5	G
5128		13	25	-43	01	7.5	G
5139		13	27	-47	29	4	GC
5194	51	13	30	47	11	8.7	G

NGC Number	Messier Number	RA h	RA m	Dec °	Dec '	Visual Magnitude	Type
5236	83	13	38	-29	52	8.0	G
5272	3	13	42	28	22	6.4	G C
5286		13	45	-51	13	7.62	G C
5457	101	14	03	54	21	9	G
5822		15	05	-54	21	6.4	O C
5866	102	15	06	55	45	11.5	G
5904	5	15	19	2	05	5.8	G C
5986		15	46	-37	47	7.12	G C
6093	80	16	17	-22	59	8.4	G C
6121	4	16	24	-26	31	7.4	G C
6171	107	16	33	-13	03	10.1	G C
6205	13	16	42	36	27	5.9	G C
6218	12	16	47	-1	57	8	G C
6231		16	54	-41	47	6	O C
6254	10	16	57	-4	06	7.6	G C
6266	62	17	01	-30	07	8.2	G C
6273	19	17	03	-26	16	8.3	G C
6341	92	17	17	43	08	6.5	G C
6333	9	17	19	-18	31	8.9	G C
6352		17	25	-47	29	6.5	G C
6397		17	41	-53	40	5.7	G C
6405	6	17	40	-32	12	4.6	O C
6402	14	17	38	-3	15	9.4	G C
IC 4665		17	46	5	43	6	O C
6441		17	50	-37	04	7.4	G C
6475	7	17	54	-34	49	3.3	O C
6494	23	17	57	-19	01	7.0	O C
6541		18	08	-43	45	6.6	G C
6543		17	59	66	37	8.6	P
6514	20	18	02	-23	02	9	H II
6523	8	18	05	-24	19	5.0	H II
6531	21	18	05	-22	30	7.2	O C
6543		17	59	66	38	8.6	P

NGC Number	Messier Number	RA h	RA m	Dec °	Dec '	Visual Magnitude	Type
6572		18	12	6	51	9	P
6603	24	18	18	-18	25	4.5	O C
6611	16	18	19	-13	46	6.0	H II
6618	17	18	21	-16	10	6.0	H II
6613	18	18	20	-17	08	8	O C
IC 4725	25	18	32	-19	14	6.2	O C
6626	28	18	25	-24	52	8.5	G C
6637	69	18	31	-32	21	8.9	G C
6656	22	18	36	-23	55	5.1	G C
IC 4756		18	39	5	29	5.1	O C
6681	70	18	43	-32	17	9	G C
6694	26	18	45	-9	24	9.3	O C
6705	11	18	51	-6	16	6.3	O C
6715	54	18	55	-30	28	8.7	G C
6720	57	18	53	31	01	9.0	P
6723		18	59	-36	41	7.32	G C
6752		19	11	-59	59	5.4	G C
6779	56	19	17	30	11	9.6	O C
6809	55	19	40	-30	57	7.1	G C
6826		19	45	50	31	8.8	P
6838	71	19	54	18	47	8.3	G C
6853	27	19	59	22	53	8.0	P
6864	75	20	06	-21	55	9.5	G C
6913	29	20	24	38	31	9.0	O C
6981	72	20	24	-12	32	10.2	G C
6992		20	56	31	43	8	SNR
7000		21	00	44	20	7	H II
7009		21	04	-11	22	8.0	P
7078	15	21	30	12	10	6.4	G C
7092	39	21	32	48	26	5.0	O C
7089	2	21	34	-0	51	6.5	G C
7099		21	41	-23	13	7.5	G C
7293		22	30	-20	51	6.5	P

NGC Number	Messier Number	RA h	RA m	Dec °	Dec '	Visual Magnitude	Type
7654	52	23	24	61	36	8.2	OC
7662		23	26	42	32	9	P